The
Changing Nature
of
Telecommunications/
Information
Infrastructure

Steering Committee on the Changing Nature
of Telecommunications/Information Infrastructure

Computer Science and Telecommunications Board

Commission on Physical Sciences, Mathematics, and Applications

National Research Council

NATIONAL ACADEMY PRESS
Washington, D.C. 1995

NOTICE: The project that is the subject of this report was approved by the Governing Board of the National Research Council, whose members are drawn from the councils of the National Academy of Sciences, the National Academy of Engineering, and the Institute of Medicine. The members of the committee responsible for the report were chosen for their special competences and with regard for appropriate balance.

This report has been reviewed by a group other than the authors according to procedures approved by a Report Review Committee consisting of members of the National Academy of Sciences, the National Academy of Engineering, and the Institute of Medicine.

The National Academy of Sciences is a private, nonprofit, self-perpetuating society of distinguished scholars engaged in scientific and engineering research, dedicated to the furtherance of science and technology and to their use for the general welfare. Upon the authority of the charter granted to it by Congress in 1863, the Academy has a mandate that requires it to advise the federal government on scientific and technical matters. Dr. Bruce Alberts is president of the National Academy of Sciences.

The National Academy of Engineering was established in 1964, under the charter of the National Academy of Sciences, as a parallel organization of outstanding engineers. It is autonomous in its administration and in the selection of its members, sharing with the National Academy of Sciences the responsibility for advising the federal government. The National Academy of Engineering also sponsors engineering programs aimed at meeting national needs, encourages education and research, and recognizes the superior achievements of engineers. Dr. Robert M. White is president of the National Academy of Engineering.

The Institute of Medicine was established in 1970 by the National Academy of Sciences to secure the services of eminent members of appropriate professions in the examination of policy matters pertaining to the health of the public. The Institute acts under the responsibility given to the National Academy of Sciences by its congressional charter to be an adviser to the federal government and, upon its own initiative, to identify issues of medical care, research, and education. Dr. Kenneth I. Shine is president of the Institute of Medicine.

The National Research Council was organized by the National Academy of Sciences in 1916 to associate the broad community of science and technology with the Academy's purposes of furthering knowledge and advising the federal government. Functioning in accordance with general policies determined by the Academy, the Council has become the principal operating agency of both the National Academy of Sciences and the National Academy of Engineering in providing services to the government, the public, and the scientific and engineering communities. The Council is administered jointly by both Academies and the Institute of Medicine. Dr. Bruce Alberts and Dr. Robert M. White are chairman and vice chairman, respectively, of the National Research Council.

Support for this project was provided by core funds of the Computer Science and Telecommunications Board. Core support for the CSTB is provided by its public and private sponsors: the Air Force Office of Scientific Research (under Contract N00014-87-J-1110); Advanced Research Projects Agency (under Contract N00014-87-J-1110), Apple Computer Corporation, Department of Energy (under Grant DE-FG05-87ER25029), Digital Equipment Corporation, Intel Corporation, International Business Machines Corporation, National Aeronautics and Space Administration (under Grant CDA-9119792), National Science Foundation (under Grant CDA-9119792), and Office of Naval Research (under Contract N00014-87-J-1110). Additional project support was provided by Ameritech Corporation, Bell Atlantic Corporation, Bell Communications Research, BellSouth Corporation, Hewlett-Packard Company, IBM Corporation, and NYNEX Corporation. Any opinions, findings, conclusions, or recommendations expressed in this volume are those of the authors and do not necessarily reflect the views of the sponsors.

Library of Congress Catalog Card Number 94-66572
International Standard Book Number 0-309-05091-X

Additional copies of this report are available from:
National Academy Press
2101 Constitution Avenue, NW
Box 285
Washington, DC 20055
800-624-6242
202-334-3313 (in the Washington Metropolitan Area)

B-457

Printed in the United States of America

iv

Preface

In 1992 the Computer Science and Telecommunications Board (CSTB) of the National Research Council (NRC) decided to convene a workshop to assess the extraordinary changes in the nation's telecommunications/information infrastructure since the divestiture of AT&T and to address related questions of technology and policy. These questions have provoked a stirring national debate about the best way to move forward in establishing a national information infrastructure. Government officials, representatives of the telecommunications and computer industries, users, and public interest groups have taken often-contradictory stands. CSTB hoped to bring to the debate the neutral and dispassionate platform offered by the NRC.

In planning the workshop, CSTB's steering committee faced two fundamental questions:

1. *What, in fact, is the telecommunications/information infrastructure?* Definitions range from raw physical connectivity, on the one hand, to connectivity plus a vast array of network and end-user services and software, on the other. It was decided not to prejudice the workshop with a limiting definition, but rather to leave the definition open to debate.

2. *On what issues should the workshop focus?* Possible foci were the exploration of potential applications as the driving force behind infrastructure development, and examination of the processes by which various players in its development can efficiently encourage and respond to new applications, no matter what they are. It was decided that the workshop's focus would be on processes, since no one can foresee which applications will prove to be important.

To make sure that the meeting was a workshop, not a series of set speeches, the committee agreed to limit attendance to about 75 people—invited from various government entities, large and small companies, and universities and "think tanks"—active in analysis or development of the telecommunications/information infrastructure. A substantial amount of time was scheduled for discussion from the floor. The workshop took place in Washington, D.C., on October 12-13, 1993, and was composed of three panels that ran consecutively. Panel 1 set the stage for the workshop by characterizing infrastructure trends and applications, while Panels 2 and 3 addressed issues of policy.

Panel 1 addressed such questions as: What is the infrastructure now, technologically and institutionally? How did it get to that state? What are the states to which it might evolve in the next decade? In connection with the last question, the needs of several large user communities—finance, health care, education, and libraries—were specifically addressed, not because these applications will necessarily drive infrastructure development but because they are a sample of the national "grand challenges" currently much in vogue. They illuminate issues of process and access underlying infrastructure development.

Panel 2 examined the role, if any, that the nation's regulatory apparatus can play in helping direct the development of the telecommunications/information infrastructure. Traditionally, the role of regulation has been to allocate resources equitably and to level the competitive playing field or tilt it toward a socially desired objective. In the past decade, the encouragement of market efficiencies has been a goal of regulation. Panel 2 grappled with the roles that regulation might play in the future. The underlying question it addressed was how regulators can simultaneously encourage infrastructure development; protect the interests of all stakeholders (including the public); and do so at social, technical, and economic costs that are substantially outweighed by the benefits.

Panel 3 considered government investment policies. The federal government has long been active in promoting growth of the telecommunications/information infrastructure through various direct and indirect means of investment: support of research and development, implementation of pilot networks, direct user subsidies, and purchasing policy. Panel 3 also debated such questions as: What is the range, in the 1990s, of public investment options? How does public investment affect market forces and are these effects desirable?

The workshop was as lively as the steering committee and CSTB had hoped. There was broad audience participation, some of which is recorded in this volume. CSTB believes that its primary goal of shedding light on fundamental issues and tensions in the national debate on the telecommunications/information infrastructure was achieved.

Thanks are due many people for their hard work and support in making the workshop a success. The steering committee laid the groundwork for the workshop, designing its content and format. A subset of committee members who became the panel chairs—Alfred Aho, Roger Noll, and David Messerschmitt (who succeeded Deborah Estrin as chair of Panel 3 when she had to give up that role)—worked assiduously in helping choose keynote speakers and panelists, in convincing them to participate, and in bringing together their respective parts of this volume. And without the determined work of the CSTB staff—Marjory Blumenthal, CSTB's director; Leslie Wade; and Renee Hawkins—nothing would have been accomplished. The anonymous reviewers prompted the steering committee and several panelists to strengthen the written record of the workshop. Finally, a special word of thanks goes to the following organizations for their financial support specifically for this project: Ameritech Corporation, Bell Atlantic Corporation, Bell Communications Research (Bellcore), BellSouth Telecommunications Inc., Hewlett-Packard Company, IBM Corporation, and NYNEX Corporation.

George L. Turin, *Chair*
Steering Committee on the Changing Nature of
Telecommunications/Information Infrastructure

Contents

PART 3—PUBLIC INVESTMENT IN
TELECOMMUNICATIONS INFRASTRUCTURE

Introduction and Overview

January 1, 1994, marked the tenth anniversary of the divestiture of AT&T. That event started an amazing period of rapid change in the telecommunications industry. Few predicted a decade ago the effects divestiture, deregulation, the exponential growth of technology, and increasingly intense international competition would have on the telecommunications/information infrastructure in the United States by 1994.

In 1984 it was quite clear what the telecommunications/information infrastructure was and who defined it. It was, in essence, the telephone and broadcast networks. The defining players were AT&T, the Federal Communications Commission (FCC), and the broadcasters. You got only the connectivity and services that were offered; compared with what is available today, it was not much.

All of this has changed radically. Instead of being defined by monopoly suppliers and regulators, the telecommunications/information infrastructure has become more closely defined by both market demand and the explosion of supporting technologies that have been brought to market by myriad suppliers. There has been much movement away from a supplier-defined infrastructure to a user- and market-defined infrastructure.

There is now an enormously complex array of networks and services, all interconnected and all changing almost daily. The distinction between broadcast and point-to-point services (e.g., in electronic mail) has blurred. So have the separations among classes of service—voice, data, video, and so on—as more services become digital. Boundaries for the production, distribution, and use of information are vanishing. Many services that were wired in 1984 are becoming wireless, and many that were wireless are becoming wired. The line between communication and computation is becoming ever fuzzier. Users are demanding ever-greater functionality, and suppliers are scrambling to respond to and lead these user demands.

All indications are that, as revolutionary as these changes are, they are just the beginning. Scarcely a day passes without the heralding of some event by the media as an indication that the day of the "information superhighway" is upon us. As we are swept along by the onrush of events, many questions arise. What, in fact, *is* the telecommunications/information infrastructure?[1] Who defines it, since AT&T and the FCC no longer uniquely do? What are reasonable roles for the various private stakeholders in the infrastructure's evolution? What should become of the government's traditional roles of regulation and various forms of infrastructure investment? On whom should the benefits arising from the new infrastructure devolve and in what proportions? Who should choose these beneficiaries? A national debate now swirls about these questions.

A Computer Science and Telecommunications Board (CSTB) workshop, titled "The Changing Nature of Telecommunications/Information Infrastructure," explored these questions. In particular, it tried to illuminate the spectrum of opinion on a broad range of potential government actions for nurturing the development of the infrastructure and to identify where consensus might be easy to achieve and where it might not. The workshop was not designed to delve deeply into the underlying technical issues, but rather to identify key technical trends and questions as they relate to economic, legal, and regulatory possibilities.[2]

This volume captures the proceedings of the workshop, building on its discussions and those of the steering committee that organized it. It contains the papers presented at the workshop, grouped into three parts reflecting the activities of the three panels and introduced by overviews prepared by the panel chairs. The papers present many different perspectives, but two distinctively different ones dominate: that of the analyst/academic and that of the practitioner/user. The former set provides the somewhat-removed sense of direction and mission important for policy development; the latter is "from the trenches," presenting the concerns of those who deal with concrete problems of application. While the gap between the two perspectives is often large, they complement each other. Policy analysts must consider the broad public interest and how to serve it. Practitioners must say what it means in practice to pursue policy goals and must identify problems and opportunities that cannot be deduced from policy principles alone.

For example, the practical experience of Citibank with networks, as explained by panelist Colin Crook, illuminates some of the prospects presented by the availability of new infrastructure. There are huge implications for the organization of economic activity (such as the ability to separate the locations for processing information from the activities generating that information) that are rarely addressed systematically in discussions of policy implications. Such insights from practice can help extend and test the more abstract analyses emerging from the policy community.

This introductory chapter summarizes the workshop discussions, capturing key panel issues and cross-cutting themes. It draws on remarks offered by participants and places their comments and the papers into perspective by referring to subsequent developments in industry and government.

PART 1—SETTING THE STAGE

Defining the telecommunications/information infrastructure is not trivial, for restricted definitions can artificially limit discussion. An appropriate conceptual framework is needed. So also is an appreciation for the scope of applications, features, and benefits desired in the modern infrastructure and for the shortcomings of the current infrastructure. Such considerations suggest the challenges faced in developing tomorrow's infrastructure, and they were the focus of Panel 1.

Charles Firestone of the Aspen Institute, one of the two keynote speakers of Panel 1, advanced a broad definition that served as a valuable foundation for the workshop's efforts and helps elucidate a number of thorny issues. The infrastructure, according to Firestone, comprises the following elements of electronic communications:

- *Production* of information in film, video, audio, text, or digital formats;
- *Distribution* media, that is, telephony, broadcasting, cable, and other electronic transmission and storage media; and
- *Reception* processes and technologies such as customer-premise equipment, videocassettes, satellite dishes, and computers.

This unconventional definition bridges both the telecommunications and information dimensions of the evolving National Information Infrastructure. According to Firestone, it "places attention on an increasingly important but often overlooked area of regulation, that of reception. As First Amendment cases move toward greater editorial autonomy by the creators, producers, and even distributors of information, attention will have to be focused on reception for filtering and literacy concerns." Conversely, as choice at the reception end is empowered—for example, as the consumer has more power to choose a telephone service provider or which subset of 500 TV channels to subscribe to—the need for government regulation of production and distribution is reduced.

According to Firestone, each of the three infrastructure components—production, distribution, and reception—has already passed through two stages of government regulation as the consumer has become more and more empowered and is now embarking on a third. An understanding of this history helps to explain why certain options are the focus of attention today, and which of these may ultimately be preferable.

Stage 1 is a regime of scarce resources (e.g., information, spectrum, equipment), centralization, monopolization, vertically dominant enterprises, and the potential for anticompetitive practices. In this stage the government's regulatory role is to control monopolies and to try to approximate the efficiencies of competition where none exist. The agenda is dominated largely by lawyers and a search for equity in the distribution of the scarce resources.

Stage 2 is a regime of abundant resources, intense competition, decentralization of control, and deregulation. The government steps back from its earlier omnipresent regulatory role, and the free market starts to replace monopolies and oligopolies. The agenda is taken over by economists and a drive for market efficiencies.

Firestone likened Stages 1 and 2 to infancy and adolescence. In the mature and complex world into which the infrastructure is heading, he believes that yet another paradigm—adulthood—is needed. This paradigm must redress the balance, as yet unachieved, among a number of cherished values: not just equity and efficiency but also liberty, community, and participatory access. The agenda in Stage 3 might be set, Firestone suggested, by political scientists and the quest for democratic access to infrastructure abundance.

Just as the conceptual framework for the infrastructure has evolved, so, too, have the technologies and the businesses that implement it. The second keynote speaker of Panel 1, Robert Lucky of Bell Communications Research (Bellcore), reviewed the past century's technological evolution of infrastructural capabilities, concentrating on the explosion of technology in the past decade. Along with that explosion has come the rapid growth of whole new industries to support it—for example, the hundreds of software firms that specialize in various aspects of the production, distribution, and reception of information. Indeed, the movement of computer-based intelligence to the periphery of the network has made some of the distinctions among production, distribution, and reception quite fluid; for instance, some components of switching—and thus distribution in Firestone's framework—are being carried out on desktop personal computers connected to the network and may now be partially identified with reception.

However, Lucky pointed out that the local loop has not changed much; it still represents a bottleneck. As Lucky put it, unlike the backbone network, there is only one subscriber on the local loop to bear its cost: "In the end, you are on your own, and there is no magic invention that makes this individual access cheap." Lucky pointed out that local access represents about 80 percent of the total investment in the (existing wireline) network and that, incrementally, it will take on the order of $25 billion to add broadband optical fiber to that access.[3] Yet there are persistent efforts to bring more capability to the local exchange and to the local loop itself. "Moreover," Lucky complained, "we are going to do it twice, or maybe even three times . . . . [That multiplicity] is called competition."

Lucky's concern about access to the local loop was shared by other workshop participants. Some worried about the effects on infrastructure development of continued monopolization of the local exchange and local loop, although new possibilities for competition in local exchange are opening up. Others concentrated on the economic, social, and technical costs that might be incurred if access to the local loop—especially the loop's extension to broadband multiple services—is not approached by regulators with extreme care. This topic is elaborated on later in this introduction.

Another key point addressed by Lucky was the contrast between the telephone network and the Internet, the former built primarily for voice and largely during Firestone's Stage 1, the latter built primarily for data (but more open in architecture) and largely during Stage 2. Lucky highlighted the contrast, noting the differing (yet evolving) pricing schemes:

> If you send a fax over the Internet, it really appears to be free. You go to the telephone on your desk, and if you send the same fax, it costs you a couple of dollars. . . . I can tell you honestly that I don't understand this. . . . I have had economists and engineers digging down, . . . [thinking] there will be bedrock down there somewhere. I cannot find it. . . . The biggest single difference . . . is a large subsidy for residential service built into the [telephone network] tariff. . . . [Internet] violates the rules of life—someone must be in charge; someone must pay.

The flat-fee charging structure associated with the Internet to date is consistent with the nature of the cost structure. Internet costs do not tend to vary with usage, nor do they have the access and settlement fees or other components found in regulated telephony.[4] The Internet experience may be instructive in examining how use of demand-side policies to stimulate network growth requires sophisticated use of the pricing mechanism so that user values can be signaled to suppliers. On the other hand, other cost elements, as Lucky suggests, may ultimately be factored in, whether to support social programs such as universal network access, or to support a growing set of interconnected networks and information resources, which may give rise to access and settlement fees.

The value of the infrastructure derives ultimately from how it is used. Four application areas examined during the workshop may drive infrastructure development: financial services, health care, education and schools, and libraries. These applications are characterized by many innovations and accomplishments but also by frustration and mismatches between desires and practical implementations. A fifth application, entertainment—much subject to mass media hyperbole but not explicitly addressed at the workshop—is also generally expected to drive infrastructure development. However, the broadband and multimedia applications associated with entertainment remain unproven, and market trials have been largely unsatisfying to both consumers and service providers.

The four application areas addressed at the workshop span a range of private and public interests. They illustrate an inexorable tendency toward international connectivity and place into bold relief the tension among Firestone's five values—equity, efficiency, liberty, community, and participatory access.

At one extreme on the scale of values are banking and financial services (described by panelist Crook), which lead in the use of modern infrastructure technology. This technology is so essential to modern banking, contended Crook, that Citibank is progressing under the "assumption that the new infrastructure for the bank essentially is a network." Efficiency is the major driving force that has led to a banking subnetwork that supports 100 billion transactions per year involving transfers of hundreds of trillions of dollars. With these dimensions, explained Crook, "there are more financial transactions in the telephone network than there are in the U.S. economy, excluding low-level cash payments." Compared to efficiency, banking places relatively little emphasis on the other four Firestone values, except for the privacy component of liberty.

At the other extreme on the values scale are applications in education, health care, and libraries. The very titles chosen for their workshop papers by panelists Robert Pearlman, a private consultant ("Can K-12 Education Drive on the Information Superhighway?"), and by Clifford Lynch of the University of California ("Future Roles of Libraries in Citizen Access to Information Resources Through the National Information Infrastructure") underscore the concerns many have about equity, community, and participatory access. As Lynch put it, "We will be challenged as a society to define a base level of information resources that we believe must be available to all members of our society, regardless of the ability to pay." This challenge was a recurring theme during the workshop and was often raised in connection with universal service, as explored further in the section "Cross-cutting Issues" below.

Considerable analysis, advocacy, and debate are being aimed, in the abstract, at demands for equity and access. Conversely, developments in such specific application areas as health care, covered by panelist Edward Shortliffe of Stanford University's School of Medicine, suggest that policy objectives in specific areas also will be important in determining the practical demands that will be made of the telecommunications/information infrastructure in the future. The "holy paradigm" for infrastructure development should be sensitive and flexible enough to respond to both policy requirements and user demands as they evolve and interact.

It is sometimes tempting to identify the information infrastructure with only the physical elements of its technology. The insights brought by panelists and attendees from their applications areas show how far off the mark that view is. As Robert Pepper of the FCC observed at the workshop,

> [W]hen we talk about infrastructure, we tend to think about wires, hardware. Infrastructure is far more than that. It is people, it is laws, it is the education to be able to use systems. If you think about the highway system, we tend to think about bridges and interstates, but the infrastructure also includes the highway laws, drivers' licenses, McDonalds along the roadside, gas stations, the people who cut the grass along the highways, and all of those support systems. You cannot talk about infrastructure in the telecom-information sector without also talking about the human support systems.

PART 2—REGULATION AND THE EMERGING TELECOMMUNICATIONS INFRASTRUCTURE

Roger Noll of Stanford University described two foci of debate about the merits of regulation. One is philosophical, considering legitimate boundaries to government coercion: Does regulatory policy go too far or not far enough in trading off individual liberty for collective rights? (Note the reprise here of the theme of values.) The other focus is policy oriented and is concerned with an instrumental question: How will the performance of a regulated industry be changed by imposing regulation? Noll suggested that even within the limited context of the historical use of regulation in the United States, the philosophical issue is unlikely to be resolved and is largely irrelevant: "Citing a meritorious policy objective is not sufficient to justify the conclusion that regulation is warranted." This conclusion reflects a recognition that regulation, while intended to correct for market failures, can itself create inefficiencies, a point made by Noll and others, including Philip Verveer of Wilkie, Farr, and Gallagher, who remarked at the workshop:

> [I]f we are interested principally in efficiency—and I think we are . . . we would be better off trying to clear out as much [regulation] as possible, recognize that there are some exceedingly legitimate issues that the local telephone industry confronts with respect to historic obligations that may no longer be sustainable or appropriate, and also that there may clearly be distributional concerns with respect to low-income people or folks who live in high-cost areas.

Consequently, maintained Noll, the debate should focus on "how to design the details of regulation to ameliorate to the maximum feasible extent the inherent infirmities of the regulatory process." He admonished that, whatever actions are taken, they must be carefully structured to be feasible within the very particular U.S. political system.

Workshop discussions during Panel 2 illustrated the extent to which we have progressed into and accepted Stage 2 of Firestone's schema: deregulation and the search for market efficiency—what Eli Noam of the Columbia Institute for Tele-Information, Columbia University, called a "post-deregulatory agenda." Indeed, there were virtually no advocates of Stage 1 regulation; the acceptance of Stage 2 was acknowledged (perhaps grudgingly) even among stakeholders, such as public interest advocates, traditionally in favor of strong regulatory positions. A consensus position seemed to be that uses of regulation have been appropriately reduced in the past decade and that in the future they should be invoked only with extreme care.

Despite relative agreement on the principle of regulatory restraint, workshop discussions reflected some of the disagreement that exists about what to do in practice; the debate on the uses and details of regulation was not one-sided. The following contrapuntal quotations illustrate the differences:

- Panelist Robert Harris of the University of California at Berkeley: "In too many instances, state regulation has become a major obstacle to competition, deployment of new technology, and development of new services. The best route to the information infrastructure of the future is not through more regulation but through different regulation and less regulation."

- Panelist Dale Hatfield of Hatfield Associates: "In either case—with limited competition (between telephone companies and cable companies) or a two-player oligopoly—the resulting rivalry would hardly meet the test for robust competition that would justify full deregulation of the local exchange carriers. Thus, policymakers and regulators would be well advised to exhibit a healthy amount of skepticism regarding the lifting of the existing line of business restrictions and deregulating local exchange carriers."

- Panelist Nina Cornell, a private consultant: "[A]lthough it has become fashionable to argue that regulation of local exchange carriers is impeding the development of new technologies, the proposed cure—deregulation—would be worse than the status quo. . . . [R]egulation, however, needs a change of focus."

- Panelist Thomas Long of the California organization Toward Utility Rate Normalization (TURN): "[R]egulators should . . . hunker down for at least 5 to 10 years of hard work . . . [in] protecting consumers and competitors from monopoly power. . . . [But] managing infrastructure improvements should not be the endeavor that keeps regulators busy in the next decade."

As Cornell also noted, decisions about regulation should be cast within a broader-than-traditional context, considering antitrust enforcement as well as classical regulatory options. This outlook was reflected by Verveer, who was lead counsel in the antitrust investigation and prosecution of AT&T. "Divestiture," he said, was itself ". . . an act of deregulation . . . [eliminating] the principal regulator . . . AT&T."

As noted in the quotations above, an issue of continuing uncertainty is timing: When is the transition to a sufficiently competitive outcome complete, allowing for further rollback of regulation? Timing has been a central concern in debates over lifting of the Modified Final Judgment constraints, encouragement or permission of entry into local exchange competition, and aspects of the universal service debate. Timing of federal regulatory decisions should also reflect actions at the state level, some of which were discussed by Harris. The states appear to have an evolving and varied role in overcoming the types of underinvestment traps described by Noam and Bridger Mitchell (then of the RAND Corporation) when new network technologies are either at their initial stages of adoption or when networks are maturing without pushing penetration to the point of achieving universal service. A function of technology development, business development, public policy, and the interaction of these factors, timing is an inherently difficult point on which to achieve consensus.

PART 3—PUBLIC INVESTMENT IN INFRASTRUCTURE

Through several forms of direct investment, the federal government, in particular, has had and can have a significant impact on the shape and growth of the telecommunications/information infrastructure. As Panel 3 keynote speaker Walter Baer of the Rand Corporation noted, federal infrastructure investments have a long history, often motivated by defense considerations. These investments go back at least to 1843, when Morse telegraphy was supported by a congressional appropriation.

As summarized by Baer during Panel 3's deliberations, investment vehicles include financial incentives (e.g., tax credits), research and development (R&D) funding, support of operating systems, development of applications, creation of information and associated resources, and support for agency activities related to standards setting. Each of these areas itself may present several options. As an example of the impact of government investment, Robert Kahn of the Corporation for National Research Initiatives examined the history of the Internet, beginning with the Department of Defense's development of the ARPANET packet-switched network, to illustrate how the federal government can involve itself in hands-on prototyping and procurement in an R&D undertaking. The government can also have less project involvement; it can undertake some form of benevolent partnering with nongovernmental entities or it can provide more hands-off oversight and steering of the enterprise. However, Baer particularly cautioned about the need for strong market feedback through industry involvement and cost sharing whenever the government's role extends beyond the R&D stage.

A strong point of agreement by Panel 3, as articulated by Baer and panelists Charles Jackson (of Strategic Policy Research Inc.) and Laura Breeden (then of FARNET), was that public investment in the telecommunications/information infrastructure is minuscule compared to private investment, in a ratio on the order of 1:50. Speaking for the Clinton administration at the workshop, Michael Nelson of the Office of Science and Technology Policy emphasized the dependence on private investment: "The administration does not have $100 billion sitting around. We are not going to build this network. We need to find incentives to encourage the private sector to spend the money that is needed." As Nelson's remarks and the discussion of regulation suggest, there are ways for the federal government to influence private spending even if it does not invest directly.

Clearly, who in industry does the investment, when, and how will depend on the environment, itself shaped by policy. Long cautioned that increasing regulated telephone rates has the effect of making ratepayers involuntary investors in advanced telecommunications infrastructure. This, he asserted, has the effect of sparing the regulated firms from competing for capital like other firms and also results in those ratepayers who do not require advanced services nonetheless helping to defray the costs of those who do. But Jackson noted that an environment that limits returns on local loop investment, for example, might drive local exchange carriers to a "harvest strategy," leaving new infrastructure construction to new entrants. A regulatory response to Jackson's concern has been a move toward price-cap rather than rate-of-return regulation, allowing carriers to increase return on investment and presumably investments by increasing their efficiency.

As discussed below, because of the rapid changes in telecommunications technologies, many think it is unlikely that regulators can accurately predict the investment behavior their policy actions will induce. This situation gives rise to the further complication, noted by Panel 2 member Robert Crandall of the Brookings Institution, that forcing (or subsidizing) investments by regulated carriers in the telecommunications/information infrastructure creates pressure on regulators to limit competition to the extent necessary to ensure a fair return on these investments. Thus, subsidized investments in new technologies can become vehicles for locking in the regulated status quo and precluding the development of competitive markets in advanced services.

The difficulty of predicting who will do what, and when, was dramatized by an event that commenced during the workshop. In the middle of the workshop, an unscheduled presentation was made announcing the proposed merger of Bell Atlantic and Tele-Communications Incorporated (a major cable TV operator), an announcement that led to much speculation and analysis in the subsequent months. The February 1994 decision by the parties not to proceed with that merger—partly on the complaint that new FCC constraints on cable TV pricing would make the merger less profitable—underscores the problems that both government and industry face in predicting and planning.

There was considerable consensus at the workshop on the need to concentrate government investment at points of market failure so as to maximize that investment's leverage. A basis for the consensus came from the conceptualization of three phases of network development by Noam of Panel 2 and was expanded on by Mitchell and Kahn during Panel 3's deliberations. (These phases should not be confused with Firestone's three *stages* of regulatory evolution.)

In the first phase, the start-up phase when the network size is small, the cost per average user is higher than the benefits accruing to that user; the network has not yet achieved "critical mass." The second phase occurs after critical mass is attained, when enough users have joined the network to reduce the average cost per user (because of the spreading of fixed costs), and the presence of more users has increased the utility (benefits) to all users. This is the regime of profitable commercial operation, where average benefits exceed average costs. As the network continues to expand, higher-marginal-cost users join, increasing the average cost while not substantially increasing network utility by their presence. These developments lead to the third phase, universal service.

Two concentration points for government investment, it was broadly agreed by workshop panelists, are phases 1 and 3. In the early stages of network development, before critical mass is attained, the government should invest to offset network start-up costs and to subsidize services (e.g., library databases) that add high utility to the infrastructure but that may not be able to bear implementation costs by themselves. It is here that government investment in fundamental research and in experimental testbeds has traditionally had significant payback, epitomized by the ARPANET and Internet experiences described by Kahn. During this period, explained Mitchell, investments can serve to lower costs, facilitating the expansion of supply, or they can serve to expand demand by generating new or improved applications. At a much later stage, phase 3, investments can be made to achieve equity and universality of access by users whose marginal cost is excessive or who cannot afford the cost of access. A number of participants argued that since even Noam's phase 2—commercial viability—has not yet been achieved in the "information superhighway," it is very premature to consider investments in universal service now. (See "Cross-cutting Issues" below.) A third concentration point for government investment, it was agreed, is standards setting, an area discussed in Kahn's paper.

Note that by focusing on the net benefit from network technology, Noam's model appears to imply that there is some way of identifying which technologies are winners and which are not, in order to target investments. Furthermore, even if all technologies are winners in some absolute social accounting sense, to the extent that they can substitute for one another, there is still the problem of picking the most beneficial. How we come by that knowledge or determine what are useful rules of thumb for making choices among alternatives is a difficult and critical question that was not well addressed by workshop participants. But this very question of targeting technologies does suggest a vital role for R&D—to evaluate options and the costs and benefits of new technologies. A recent NRC report suggests that it is possible to encourage development and deployment of the information infrastructure in ways that leave open choices of implementing technologies (CSTB, 1994). When government support extends beyond the R&D stage, however, it is important to assess and respond to market feedback.

Despite the cautions expressed above, several workshop participants were passionate in their conviction that federal investments in infrastructure-related R&D have been and will continue to be enormously productive. The skepticism expressed on this point by others is in part symptomatic of the broader fiscal context in which federal investment in R&D is being reexamined. Because of changing constraints on government investment, Kahn speculated that the federal government probably could not repeat what it had done in launching the National Research and Education Network (NREN) program, which built on and fueled development of the Internet. That is, although the Clinton administration's National Information Infrastructure (NII) initiative is sometimes perceived as involving both substantial government investment and control, neither is realistic. Kahn posed the question of how the process that resulted in NREN and its successes could be institutionalized as part of the R&D business. The larger question is how the Internet model can be leveraged to advance the NII.[5]

CROSS-CUTTING ISSUES

Although policymakers face choices among various options, some of which can be grouped under the headings of regulation or public investment, several issues cut across policy regimes. Such issues were raised by all three panels and include accommodating rapid technology change, standards and standards-setting, and democratization of the infrastructure. Verveer listed a similar set of issues: interconnection and standards, open entry to the local exchange, universal service, and telephone company integration into adjacent markets.

Verveer's last issue relates to the others as a factor affecting timing and direction, because the telephone network (and the cable networks as well) were originally built for efficient delivery of one service—voice communications—whereas the emerging concept of the NII implies an architecture or technical framework that is more general and flexible, and thus more capable of supporting multiple services (CSTB, 1994). Technological changes and standards will be prerequisites; the nature of the technology and architecture, in turn, will affect who benefits from the infrastructure and when, as well as what, policy intervention(s) will be most effective.

Accommodating Rapid Technological Changes

Rapidly changing technology poses major challenges for infrastructure policy. The rapid shift in the telephone network system from analog to digital switching technology, described by Lucky, was but the first of several waves of change that are transforming, integrating, and adding complexity to the telecommunications/information infrastructure. The rapid change (and uncertainty) is today underscored by the contrasts—previously noted—between the telephone network and the Internet. As Lucky and Kahn explained, these networks involve very different uses of technology and also very different technical architectures, and their cost structures and patterns of use are also very different. Complicating the situation is the fact that existing understanding and models were conditioned on the telephone infrastructure, whereas different technologies and architectures—such as the Internet approach toward networking, wireless transmission, and competition from or interconnection with cable networks—will change many assumptions (e.g., about urban-rural cost differentials).

Furthermore, the emphasis on intelligence at the periphery of the network, which is characteristic of the Internet, will have implications for the logic of network operations, its structure, and especially the types of public infrastructure services that will be in demand in the future. Conversely, the need for peripheral intelligence also highlights problems associated with variations in ability to pay. For example, Pearlman's discussion of conditions in education shows how limited resources for communication outside the traditional public information infrastructure, in this case computers and telecommunications facilities in public schools, constrain the use of public infrastructure to promote educational policy objectives.

Both sets of concerns—network operations and affordability—will be affected by decisions about network architecture. The openness of an Internet-style architecture, for example, implies a movement toward the unbundling of network facilities and services, which will, in turn, affect costs, competition, and innovations in both fundamental technologies and the services that build on them (CSTB, 1994).

The rapidity of technological change is also reflected—along with a certain amount of media "hype"—in the proposed broadband multimedia applications of the NII. While typical Internet applications (e.g., electronic mail, file transfer) and other narrowband applications are now available with no more than the current twisted-pair telephone connections, implementation of a broadband NII has enormous technical, economic, and policy implications.[6] It will require billions of dollars, but so far—some at the workshop asserted—its applications are unproven. The skeptics therefore questioned the wisdom of the pell-mell dash to an advanced-feature NII. In response, Kahn, one of the primary architects of the Internet (via its predecessor, ARPANET), replied that the original plans for ARPANET provoked the same sort of skepticism from the research community, the very user set that initially benefited most from it. The skepticism dissolved only when an ARPANET testbed was available. Kahn asserted that, until a committed user community has developed for the advanced-feature NII, skepticism toward it will remain. He felt strongly that a government-sponsored testbed is the way to develop such a user community.

While infrastructure investment discussions typically focus on hardware elements, substantial portions of the emerging telecommunications/information infrastructure consist of software —amounting to billions of dollars of investment. Software will become increasingly important because it is what makes possible the development of common (and differentiated) infrastructural *services*, which greatly enhance the usefulness of the physical components of the infrastructure. The spur to the growth in Internet usage resulting from the introduction of the Mosaic interface to the World Wide Web illustrates the impact of software. But one often unappreciated problem, noted by Alfred Aho, then of Bellcore, is that substantial amounts of embedded software present a bottleneck to change. Another problem, noted by Baer, is a tax structure that sometimes makes it difficult to reward or encourage investments in developing new kinds of software; for example, federal research and experimentation tax credits in place since 1981 have not always been interpreted to be applicable to software development.

The rise of software as a vehicle for adding value to underlying network services has enabled the growth of network-based systems integration. Noam, Breeden, and others remarked on the competition between network service providers and systems integrators. Reflecting on the evolving patterns of competition, several participants speculated about which capabilities and services could or would be regarded as commodities and what commoditization may mean for technology development and deployment as well as the evolution of the industry.

The criticisms leveled at regulation during the Panel 2 session reflected an appreciation for the association between past regulation and the relatively slow innovation and modernization in telecommunications. Part of the problem is a human one, the difficulty faced by regulators trying to decide which technologies are preferable. Observed Crandall,

> [T]he technology is changing so quickly in this area and the number of players is so potentially large that trying to use regulation as the instrument for inducing investment in infrastructure is fraught with enormous dangers. Even if regulation could

bring about the type of investment that you think is desirable today, it would be inefficient to do it through regulation, it would be extremely costly, and what I fear is if it turned out to be a mistake, it would be very difficult to correct that mistake.

Crandall's concerns were complemented by Lynch's observations on how information technologies are inducing changes in the complex system of information providers of which libraries are a part. The process of change, apparent in citation and publishing practices, is ongoing. Neither the end results nor their sensitivity to coercive policy interventions, such as regulation, can be forecast.

Concerns about the risks of the slow pace of regulatory processes and the costs of mistakes imply assumptions that (1) the more rapidly technology advances the better and (2) there will be no cause for regret over the outcomes if we just let natural economic forces take their course. Both assumptions call for further analysis. Indeed, a CSTB committee that considered these issues has argued that market forces may not readily yield either the unifying, open architecture or the kinds of general and flexible technology required to maximize the societal benefits implied by much of the NII rhetoric (CSTB, 1994). To minimize mistakes, commented Cornell, participation by multiple players should be encouraged "so that lots of new ideas can be tried in the marketplace." The opening of the local exchange to competition is consistent with this view, although the proliferation of alliances across industries (e.g., between telephone and cable or communications and software or hardware systems companies) suggests new possibilities for reducing healthy competition among approaches. Also, as Hatfield observed, competition will be affected by the structural and human knowledge constraints on cable companies providing telephone service and on telephone companies providing cable service.

All this having been said, there was broad consensus on the poor fit between the political process by which regulations are made and that of innovation. As Noll noted, "Technological progress is not something that happens through consensus and compromise. It happens with crazy people going out and doing things that others did not think could be done." The fact that much current and anticipated technological change is not incremental underscores the weaknesses of "the incremental consensus approach of regulation and policymaking," according to Pepper, who suggested that "there may be an opportunity to have some nonincremental change on the policy side."

Standards and Standards Setting

Standards are essential for making a multifaceted multiprovider infrastructure work. They can shape the nature of the services and capabilities that are available, as well as how they are implemented (see Box 1). Although, as noted above, many comments were made at the workshop about the difficulties and therefore the undesirability of having bureaucrats select specific technologies or players, Pepper noted that that mind-set should not preclude a major government role in the standards arena:

[T]here is a very important role for the government to establish the framework within which the smartest engineers in industry develop standards. . . . [T]here is a balancing because people . . . who might lose in a standards-setting process too often try to use the process against the setting of standards that might be somebody else's standard.

BOX 1 The Economic Role of Standards

Joseph Farrell
University of California at Berkeley

Standards are an explicit or implicit agreement to do things—such as encode information—in common ways so that different services can work together, information can be exchanged easily, and so forth. Standards may be dictated, benevolently or not, by an authority or powerful player (often a seller or buyer). This was historically the main mode of standardization in telecommunications, with AT&T in the United States and governments elsewhere determining the standards. But as the telecommunications industry has become less concentrated, authority-dictated standards are no longer available and other means of reaching agreement must be developed instead. These methods include the use of a formal or informal consensus process ("committees," or voluntary standards organizations) and a process of letting "de facto" standards emerge from a positive-feedback process among many industry participants.

None of these mechanisms for generating standards always generates a "good" standard, a rapidly available standard, or any standard at all. Government-set standards may be unduly influenced by pressure groups or simply reflect a lack of sophistication in underfunded government agencies, and standards set by big players may be chosen to preserve their position. Consensus negotiations may be protracted, and the outcomes may reflect bargaining power more than technical or economic merit. Finally, the economic forces that usually tend to make de facto marketplace outcomes economically efficient are at best weak in contexts where standards are important, as the recent economic literature on the subject has shown.

Part of the problem with standards development is that standards not only make interconnection and information exchange easier, but they also affect the nature of competition. Open standards level the playing field, reducing the commercial advantages conferred by size or historical dominance. They also facilitate specialization, especially specialized entry. These complex competitive effects often give participants in standards-setting efforts mixed motives, or even create incentives for sabotage.

Even if all participants want a standard, problems remain. Perhaps likelier than a choice of "the wrong" standard is the wrong timing of the standards decision: a choice of standard may freeze the fundamental technology (although it may accelerate nonfundamental innovation), and so the timing of standardization requires a subtle balancing of the benefits with the foreclosure of options. With hindsight, the Japanese and European governments probably chose high-definition television (HDTV) standards prematurely, slightly too soon for full digitization, whereas the United States (fortuitously, and by a hair's breadth) probably waited long enough. Some would say that the same is true of digital wireless telephone. But without hindsight, the timing problem is hard: opinions will genuinely differ on the payoff for waiting (most experts in 1990 thought that all-digital HDTV within the confines dictated by the Federal Communications Commission (FCC) was not possible, until General Instrument showed otherwise), and people experience different trade-offs between speed of standardization and the technical quality of the standard chosen. Moreover, if a market is developing in advance of standardization, participants may become locked in and standards deliberations become moot. Many standards organizations, including the International Telecommunications Union and the International Electrotechnical Commission, have begun to pay more attention to the speed and timeliness of their standards development, which should reduce unintended delays but will probably do nothing for misjudgments about optimal timing.

In the United States, "government" standards setting often is closer to government-overseen standards setting. For instance, in HDTV, the FCC created the Advisory Committee on Advanced Television Service, involving industry participants, and asked it to recommend a standard. This approach echoes the FCC's approach with earlier generations of television standards: indeed, the current television "NTSC" standards are named for the National Television Systems Committee. Compared to European governments, for instance, the U.S. government has also been quite reluctant to set standards.

continues

BOX 1—*continued*

Government involvement may be less risky in standards setting than in some other forms of regulation, because even standards set with government involvement are often voluntary and can be ignored if unsuitable enough: if there is a clearly superior alternative it may replace a poorly chosen standard. For example, the U.S. government tried to use its purchasing power to encourage the use of Open Systems Interconnection (OSI) computer networking protocols, but attractive products were not readily forthcoming, and most other buyers continued to prefer the more established TCP/IP protocol suite. As a result, the government recently withdrew its procurement specification of OSI.

Of course, this is at best an imperfect check on errors—the government's choices are powerful and can impose bad outcomes. But government involvement in standards, if properly managed, may be more like "indicative planning" and less like compulsory regulation. This can potentially help private parties coordinate without coercion.

Economics Literature on Standardization

Economists have long recognized that some products are more valuable when more people use them; the phenomenon we now call "network effects" was studied many years ago by Leibenstein (1950). But the modern literature took off when economists became interested in how firms might make strategic choices in the arena of standards—particularly whether to try to become compatible with rivals or insist on remaining incompatible: for recent discussions, see Katz and Shapiro (1994) and Besen and Farrell (1994). Further references can be found in these latter two articles and in a survey by David and Greenstein (1990).

In characterizing the FCC's involvement in standards for high-definition television (HDTV), Pepper attested to the role of functional standards in militating against overly short-term technical solutions. Similarly, Kahn's remarks addressed how government funding of Internet protocol development and standardization resulted in an architecture that embraces heterogeneity (i.e., heterogenous hardware, software, underlying transmission technologies, and capacities).

Several workshop participants commented on elements of the standards-setting process that they believed should be promoted. Lucy Richards of the Department of Commerce, for example, commended the FCC for pushing the HDTV standards process to involve not just broadcasters but also the computer industry and small innovative technology firms. Vinton Cerf of MCI, drawing on his experiences with the Internet, noted that interoperability standards must involve a process requiring implementation and testing that shows that independent implementations can, indeed, interoperate.

Although they have received the most attention in discussions of information infrastructure, interoperability standards are only one of the many kinds of standards needed. Another kind of standard relates to the encoding and formatting of the information to be transmitted, stored, and accessed via the infrastructure. Comments made by Shortliffe describing difficulties in the health care context appear to apply more generally; they are relevant to the development of applications in several areas. Shortliffe noted, for example, that because it is difficult to encode electronically subjective knowledge and impressions about quality of care that are important in making informed medical and health insurance decisions, there may be a danger that reliance on electronic formats will exclude useful data and reduce efficiency. There is thus a need for more work on information classification or categorization systems, which should properly be considered part of the infrastructure. Shortliffe commented on the standards-like nature of such classification systems, noting the need to eliminate inconsistencies across information classification or categorization systems that impede effective use of data already collected. His observations from health care were echoed by the more general assertions by William Gillis of Motorola that investing in a uniform organization and structure for data would make access to and use of data more efficient and could ultimately lower the costs of accessing and publishing data.

A wag once pinpointed the beauty of standards: everyone can have his own. The irony identifies the tension in standards setting. Were governments to have the authority and responsibility for setting unique, definitive standards, one could easily have bureaucratic infighting leading to ossification of standards while technology leapfrogs them—witness the endless international bickering over narrowband integrated services digital network (ISDN) standards while ISDN became all but obsolete, or the federal promotion of the Open Systems Interconnection (OSI) scheme through GOSIP while markets moved toward TCP/IP in data network technology. On the other hand, if companies compete to set proprietary standards to gain market share, chaos may be the result. A popular recent middle ground—establishment of multicompany consortia, sometimes subsidized by government, to create de facto industry standards—has been attacked by larger companies on the grounds of inefficiency and by smaller companies on the grounds of exclusivity. Workshop participants seemed to agree that the frequently referenced government roles in fostering the TCP/IP standards through support of Internet development and use (i.e., setting standards by supporting testbed development and research projects) and the HDTV standard through competitive testing for spectrum allocation (i.e., setting standards by forcing competing interests to cooperate) may be models for future government action.

Democratization

Many workshop participants, beginning with Firestone, appeared to envision a greater democratization of information access and use. This theme was explored in discussions of universal service, the changing roles of such information providers as libraries, and increasing access and use in education. The Internet was at least implicitly recognized as a model of democratization, since it has demonstrated the ability of small players to become providers of content to a global network market. On the Internet, even individuals can make Motion Picture Experts Group (MPEG)-encoded movies available to large numbers of people; no longer is movie distribution limited to the large studios. These features of the Internet reflect its technology and architecture and therefore cannot be assumed to translate automatically to the larger, more complex NII.

The fact that existing policy mechanisms may be ill suited to promote democratization was recognized. As Pepper observed, "[T]he traditional regulatory process all too often puts agencies like the FCC in the business of refereeing between the private interests of providers as opposed to putting users and user needs at the forefront." Industry faces similar problems. Crook, for example, lamented that the "most complicated part of networking . . . is how to deliver things to ordinary people who walk in off the street."

Education represents an applications area that is fundamental to the democratization of infrastructure access and use—and one that epitomizes the challenge of achieving that democratization. Difficulties experienced in integrating the information infrastructure into education underscore the importance of the human and institutional factors needed to realize the potential benefits promised by the hardware and software components. As Pearlman recounted, there are already many examples of K-12 schools obtaining access to events, people, and information around the world, sometimes in real time. However, since most of the enabling services and resources are underwritten by parties other than the educators or students, "one of the characteristics that unites all these wonderful examples is that most have truly demonstrated their commercial unavailability," Pearlman pointed out.

Another route for using public institutions to democratize access to information and the information infrastructure is public libraries, which, as noted by Carol Henderson of the American Library Association, serve as community institutions and information providers. Workshop participants suggested, however, that no single role was likely to emerge for libraries. Reflecting on the capabilities made possible by technology, for example, Linda Roberts of the Department of Education speculated that, with access to the information infrastructure, individuals could have libraries in their own homes. Moreover, based on experience in education, such personal libraries would not necessarily contain information derived only from the publishing industry—that is, not only might the role of libraries change, but so, too, might the role of other information providers, in part because individual infrastructure users are likely to be information providers as well as consumers.

The many possible ways in which libraries could change were addressed by several speakers and summarized by Lynch:

> [I]t is important to recognize that libraries are part of a much more complex system of information providers and consumers which includes publishers; the government, which creates a certain amount of public domain information; scholars who use it; [and] researchers. . . . At the same time libraries are struggling to modernize into an environment where much of their content is electronic. Information technology in networks is going to cause some fundamental reassessment of the roles of other groups within which libraries are part of the public system.

Fundamental to democratization is the achievement of universal service, which will be shaped by regulation—driving private investment—and public investment. George Gilder, a private consultant, cautioned that "you cannot instantly create universal service," but the political and moral imperatives of striving for it pervaded the workshop discussions. Universal service has been advanced as an essential principle by the Clinton administration, Congress, and a range of consumer, industry, and public interest advocacy groups; it was also the focus of a set of commentaries commissioned by the administration in October 1993 and published in March 1994 (NTIA, 1994). Despite a tendency in the public debate to polarize the possibilities with respect to those who have access and those who do not, there will clearly be a spectrum of possible outcomes, which may itself evolve over time.

The recognition that network evolution passes through stages and achieves critical mass over time suggests that universality may naturally be delayed. As Noam explained, "There is a connection between critical mass and universal service. And that is you would never get to that second problem without having resolved the first problem." Moreover, although competition may contribute to network expansion, Noam cautioned against assuming it would solve the universal service problem—competition can help make production of network services more efficient, but that issue is different from allocatable distribution. Competition can also complicate the challenge of collecting funds for universal service subsidies if the current model of collection by service providers is presumed to continue. Alternative models for collecting and distributing communications-related subsidies may be needed.

Two main dimensions of the universal-service problem were discussed by workshop participants. One, clearly the more general concern, is access by individuals with low incomes regardless of where they live. Workshop participants suggested that a critical factor in terms of cost and timing would be the choice of mechanism—some vehicles for extending service to individuals who could not afford it on their own are more efficient than others; the fact that one vehicle has been used in the past should not be the basis for selecting a vehicle in the future.

The other dimension, on which a greater range of opinions was expressed at the workshop, is access in rural areas. Some skepticism surfaced about whether the historic rationale for subsidizing rural access can be justified. Commented Crandall, "[T]here is no particular reason at this stage of our economic development to suggest that people who live in the country are so poor that they could not pay for the full cost of their telephone service."

The tension among the values articulated by Firestone (equity, efficiency, liberty, community, and participatory access) was expressed in the range of opinions on the roles for private and public action in achieving democratization. For example, some felt that if infrastructure access were as important to schools, libraries, and health care organizations as it is to banks, these organizations would rationally reallocate their resources to buy such access at the expense of other needs. Others expressed the belief that schools and libraries, if left to their own resources, would never have the wherewithal to buy the access they need; lacking government intervention, they claimed, a large information-disadvantaged class will arise. The confounding factor of politics also was invoked as a constraint on access and use of the information infrastructure. Commented Noll,

> [I]f the world is such that the politics of education select against the delivery of information to students in favor of something else, it is exactly the same politics that will drive a regulatory institution. They are not separate. They are run by the same state legislature and governor.

CONCLUSION

The decade since the divestiture of AT&T and the ensuing deregulation have wrought huge changes. In reflecting on these changes, Verveer (a key player in the 1984 events) characterized some of these changes (Box 2). The telecommunications/information infrastructure has become a significant and increasingly important component of the social overhead that contributes to the productivity and growth of other economic activities. This social overhead factor is one reason for treating the development of policies for telecommunications and information industries as somehow different from the development of policies for other industries.

The issues discussed at the workshop illuminate several possible future roles for government. Pepper identified six government roles:

1. Goal-setting and leadership;
2. Setting the regulatory framework, which provides incentives for investment;
3. Facilitating or establishing the framework for the standards-setting process;
4. Reducing risk at the margins (e.g., by supporting R&D and developing innovative applications);
5. Procurement, which provides incentives for investment; and
6. Addressing market failures on the communications and information sides.

The breadth of opinions aired at the workshop shows that decisions on each possible government role will not be easy. As Michael Roberts of EDUCOM remarked, "Certainly, this [workshop] has shown how primitive the policy space is around the NII. We have to manage that issue, not wish hopefully that it will go away." The failure of the 103d Congress to pass telecommunications reform legislation underscores the value of more fully understanding the issues and perspectives discussed in this report.

In short, the workshop elicited lively and at times contentious discussion of key issues of governmental action in the development of the telecommunications/information infrastructure. A surprising degree of consensus on some of the immediate issues emerged. Some of these points of consensus are as follows:

BOX 2 Lessons from the Postdivestiture Decade

Philip Verveer
Wilkie, Farr, and Gallagher

1. Competition and its counterpart, deregulation, work.

2. The Modified Final Judgment's (MFJ) institutional structure is fragile but sustainable as long as the Department of Justice lends material support or the presiding judge is a person of exceptional ability and background.

3. Realignment of the comparative authority of federal and state regulatory authorities is needed in the wake of the Supreme Court decision in *Louisiana Public Service Commission v. Federal Communications Commission*, 476 US 355 (1986).

4. The salient question about the line of business restrictions is not if they will be removed but when.

5. The divestiture and arrangements that surround the MFJ, including the line of business restrictions, made way for technological and consumer demands, just as the cast in the AT&T case believed they would.

• Government spending on the NII, as large as it is, is still a very small percentage of total spending on the information infrastructure. The government should therefore not be thinking about building the NII, but rather about creating conditions that promote private-sector investment. Key leverage areas are precompetitive R&D (in which a free market tends to underinvest) and regulatory policy.

• The deregulatory trend of the past two decades has been basically healthy and should be continued, although with some caution; there are still roles for regulators, but these are much changed.

• Regulation—particularly state regulation—has often impeded the adoption of innovative technologies. Given the exponential advance of technology, incremental changes in regulation in the future could be a major stumbling block to implementation of the NII.

• Universal service is a long-term goal that must assist the NII's achievement of critical mass and profitability.

• As emphasized by the Clinton administration's NII initiative, potential applications in the areas of education, health care, and libraries could serve valuable social goals, but it must be recognized that:

  • K-12 schools form a "cottage industry" that is woefully unprepared to utilize an NII; most classrooms lack even the most rudimentary high-technology tools, including access to telephone lines; and

  • Health care providers are typically a decade or more behind in the use of information technology.

- Electronic distribution of information may so greatly change the economics and other aspects of publishing and libraries that the future forms of these institutions are unpredictable.

Fundamental philosophical differences never lie far from the surface, however, indicating that the current debate over development of the telecommunications/information infrastructure will not soon die out. Some of these differences are as follows:

- There are those who think that our inability to predict the course and outcome of technological developments severely restricts the utility of any governmental intervention—regulation or investment—in market processes and those who believe that we know enough about the general patterns by which infrastructure industries develop to make useful policy prescriptions.

- Even within the consensus, noted above, that government investment in precompetitive R&D for the NII will continue to be productive, a few workshop participants expressed skepticism that anything like the success of the Advanced Research Projects Agency in developing the seeds of the Internet could be repeated.

- There are those who would immediately reduce regulation to the vestigial areas where technological abundance has not been achieved (e.g., the local exchange) or eliminate it altogether, trusting market efficiencies. Others argue that regulators have decades of work ahead of them (e.g., in setting terms of interconnection, allocating costs) before they perish in the full blaze of Firestone's Stage 2.

- There are those who abhor any government intervention in the standards-setting process, claiming that market mechanisms ("shakeouts") are more efficient, and there are those who insist that, without a government role, interoperability and compatibility will suffer greatly.

The persistence of disagreement on these and other issues, and the related likelihood that policy reform related to telecommunications/information infrastructure will be an ongoing process, suggest that such issues are fruitful fodder for research as well as further debate.

The papers presented in this volume, as well as the introductory and discussion material for Parts 1 through 3, provide perspectives on potential roles and positions for the federal government, illuminating the pros, cons, and trade-offs.

REFERENCES

Besen, Stanley M., and Joseph Farrell. 1994. "Choosing How to Compete: Strategies and Tactics in Standardization," *Journal of Economic Perspectives* 8(2):117-131.

Computer Science and Telecommunications Board (CSTB), National Research Council. 1994. *Realizing the Information Future: The Internet and Beyond.* National Academy Press, Washington, D.C.

David, Paul, and Shane Greenstein. 1990. "The Economics of Compatibility Standards: An Introduction to Recent Research," *Economics of Innovation and New Technology*, Vol. 1. Harwood Academic, Chur, New York.

Katz, Michael L., Carl Shapiro. 1994. "Systems Competition and Network Effects," *Journal of Economic Perspectives* 8(2):93-115.

Leibenstein, Harvey. 1950. "Bandwagon, Snob, and Veblen Effects in the Theory of Consumers' Demand," *Quarterly Journal of Economics,* Vol. 64.

National Telecommunications and Information Administration (NTIA). 1994. *20/20 Vision: The Development of a National Information Infrastructure.* U.S. Department of Commerce, Washington, D.C.

NOTES

1. This report uses the term "telecommunication/information" because the facilities and concepts associated historically with the telecommunications infrastructure and more recently with the information infrastructure are fundamentally intertwined. That linkage, in fact, is fundamental to much of the discussion captured in this report.

2. A recent CSTB report, *Realizing the Information Future* (CSTB, 1994), addresses many relevant technical issues.

3. This represents the best estimate available at the time, recognizing that the costs of upgrading the telephone network from copper to optical fiber have been falling.

4. See Chapter 5 in CSTB (1994).

5. For a discussion of related issues, see CSTB (1994).

6. Of course, there are broadband NII capabilities today, most notably in cable television systems. As typically discussed, references to "a broadband NII" embrace a bigger mix of services, greater two-way service, and more integration of services than has been typical of cable television offerings.

Part 1

Setting the Stage

Introduction to Part 1

Alfred V. Aho

The telecommunications/information infrastructure in the United States has been evolving steadily for more than a century. Today, sweeping changes are taking place in the underlying technology, in the structure of the industry, and in how people are using the new infrastructure to redefine how the nation's business is conducted.

The two keynote papers in this section of the report set the stage by outlining the salient features of the present infrastructure and examining the forces that have led to the technology, industry, and regulatory policies that we have today.

Robert Lucky discusses how the nation's communications infrastructure evolved technically to its current form. He outlines the major scientific and engineering developments in the evolution of the public switched telecommunications network and looks at the current technical and business forces shaping tomorrow's network. Of particular interest are Lucky's comments contrasting the legislative and regulatory policies that have guided the creation of today's telephone network with the rather chaotic policies of the popular and rapidly growing Internet.

Charles Firestone outlines the regulatory paradigms that have molded the current information infrastructure and goes on to suggest why these paradigms might be inadequate for tomorrow's infrastructure. He looks at the current infrastructure from three perspectives—the production, electronic distribution, and reception of information—and proposes broad goals based on democratic values for the regulatory policies of the different segments of the new information infrastructure.

The four papers that follow examine the use of the telecommunications infrastructure in several key application areas, identify major obstacles to the fullest use of the emerging infrastructure, and discuss how the infrastructure and concomitant regulatory policies need to evolve to maximize the benefits to the nation.

Colin Crook discusses how the banking and financial services industries rely on the telecommunications infrastructure to serve their customers on a global basis. He notes that immense sums of money are moved electronically on a daily basis around the world and that the banking industry cannot survive without a reliable worldwide communications network. He underscores the importance of an advanced public telecommunications/information infrastructure to the nation's continued economic growth and global competitiveness.

Edward Shortliffe notes that the use of computers and communications is not as advanced in health care as in the banking industry. He gives examples of how the use of information technology can both enhance the quality of health care and reduce waste. He stresses the importance of demonstration projects to help prove the cost-effectiveness and benefits of the new technology to the health care industry.

Robert Pearlman notes that, at present, K-12 education lacks an effective information infrastructure at all levels—national, state, school district, and school site. He presents numerous examples of how new learning activities and educational services on an information superhighway have the potential for improving education. He outlines the major barriers that need to be overcome to create a suitable information infrastructure for effective K-12 schooling in the 21st century.

In the final paper, Clifford Lynch looks at the future role of libraries in providing access to information resources via a national information infrastructure. He examines the benefits and barriers to universal access to electronic information. He notes that a ubiquitous information infrastructure will cause major changes in the entire publishing industry and that intellectual property rights to information remain a major unresolved issue.

The Evolution of the Telecommunications Infrastructure

Robert L. Lucky

I have been asked to talk about the telecommunications infrastructure—how we got here, where we are, and where we are going. I don't think I am going to talk quite so much about where we are going but rather about the problems. I will discuss what stops us from going further, and then I will make some observations about the future networking environment.

First, I will review the history of how we got where we are in the digitization of telecommunications today, and then I would like to address two separate issues that focus on the problems and the opportunities in the infrastructure today. The first issue is the bottleneck in local loop access, which is where I think the challenge really is. Then I will discuss two forces that together are changing the paradigm for communications—what I think of as the "packetizing" of communications. The two forces that are doing this are asynchronous transfer mode (ATM) in the telecommunications community and the Internet from the computer community. The Internet will be the focus of many of the subsequent talks in this session, since it seems to be the building block for the national information infrastructure (NII).

HOW THE TELECOMMUNICATIONS NETWORK BECAME DIGITAL

It is not as if someone decided that there should be an NII, and it has taken 100 years to build it. The history of the NII is quite a tangled story.

First, there was the Big Bang that created the universe, and then Bell invented the telephone. If you read Bell's patent, it actually says that you will use a voltage proportional to the air pressure in speech. Speech is, after all, analog, so it makes a great deal of sense that you need an analog signal to carry it. In the end Bell's patent is all about analog. Since that invention, it has taken us over 100 years to get away from the idea of analog.

A long time went by and the country became wired. In 1939, Reeves of ITT invented pulse code modulation (PCM), but it was 22 years before it became a part of the telephone plant, because nobody understood why it was a good thing to do. Even in the early 1960s a lot of people did not understand why digitizing something that was inherently analog was a good idea. No one had thought of an information infrastructure. Computer communications was not a big deal. The world was run by voice. But since then the telephone network has been digitized. As it happened, this was accomplished for the purposes of voice, not for computer communication.

So in 1960 we had analog voice and it fit in a 4-kilohertz channel, and we stacked about 12 of these together like AM radio and sent it over an open wire. That was the way communication was done. But if you take this 4-kilohertz voice channel and digitize it, it is 10 times bigger in bandwidth. Why do we do that? That seems like a really dumb idea. But you gain something, and that is the renewability of the digital form. When it is analog and you accumulate distortion and noise comes along, it is like Humpty Dumpty—you can't put it back together again. You suffer these degradations quietly.

The digits can always be reconstructed, and so in exchange for widening the bandwidth, you get the ability to renew it, to regenerate it, so that you do not have to go 1,000 miles while trying to keep the distortion and noise manageable. You only have to go about 1 mile, and then you can regenerate it—clean it up and start afresh. So that was the whole concept of PCM, this periodic regeneration. It takes a lot more bandwidth, but now you can stack a lot more voice signals, because you don't have to go very far.

The reason that the network is transformed is not necessarily to make it more capable but to make it cheaper. PCM made transmission less expensive, since 24 voice channels could be carried, whereas in the previous analog systems only 6 voice channels could be carried. So digital carriers went into metropolitan areas starting in the early 1960s.

At that time we had digital carriers starting to link the analog switches in the bowels of the network. More and more the situation was that digits were coming into switches designed to switch analog signals. It was necessary to change the digits to analog in order to switch them. Engineers were skeptical. Why not just reshuttle the bits around to switch them?

THE ADVENT OF DIGITAL SWITCHING

So the first digital switches were designed. The electronic switching system (ESS) number four went out into the center of the network where there were lots of bits and all you had to do was time shifting to effect switching. That seemed natural because all the inputs were digital anyway. It was some time before we got around to the idea that maybe the local switches could be digital, too, because the problem in the local switches is that, unlike the tandem switches, they have essentially all of their input signals in analog form.

I was personally working on digital switching in 1976. I did not start that work, but I was put in charge of it at that time. I remember a meeting in 1977 with the vice president in charge of switching development at AT&T. It was a very memorable meeting: we were going to sell him the idea that the next local switch should be digital.

We demonstrated a research prototype of a digital local switch. We tried to explain why the next switch in development should be digital like ours, but we failed. They said to us that anything we could do with digits, they could do with analog and it would be cheaper—and they were right on both scores. Where we were all wrong was that it was going to *become* cheaper to do it digital, and if we had had the foresight we would have seen that intelligence and processing would be getting cheaper and cheaper in the future. But even though we all knew intellectually that transistor costs were steadily shrinking, we failed to realize the impact that this would have on our products. Only a few short years later those digital local switches did everything the analog switches did, and they did more, and they did it more cheaply.

The engineers who developed the analog switches pointed to their little mechanical relays. They made those by the millions for pennies apiece, and those relays were able to switch an entire analog channel. Why would anyone change the signals to digits? So the development of analog switching went ahead at that time, but what happened quickly was the advent of competition enabled by the new digital technology. There was a window of opportunity where the competition could come in and build digital switches, so AT&T was soon forced to build a digital switch of its own, even though it had a big factory that made those nice little mechanical relays.

The next major event was that fiber came along in 1981. Optical fiber was inherently digital, in the sense that you really could not at that time send analog signals over optical fiber without significant distortion. Cable TV is doing it now, but in 1980 we did not know how to send analog signals without getting cross-talk between channels. So from the start optical fiber was considered digital, and we began putting in fiber systems because they were cheaper. You could go further without regenerating signals than was possible with other transmission media, so even though the huge capacity of optical systems was not needed at first, it went in because it was a cheaper system. It went in quickly beginning in 1981, and in the late 1980s AT&T wrote off its entire analog plant. The network was declared to be digital.

This was incredible because in 1984—at divestiture—we at AT&T believed that nobody could challenge us. It had taken us 100 years to build the telephone plant the way it was. Who could duplicate that? But what happened was that in the next few years AT&T built a whole new network, and so did at least two other companies! And, in fact, just a few years ago, you would not have guessed that the company with the third most fiber in the country was a gas company—Wiltel, or the Williams Gas Company. So everybody could build a network. All of a sudden it was cheap to build a long-distance network and a digital one, inherently digital because of the fiber.

In the area of digital networking we have been working for many years on integrated services digital network (ISDN). We all trade stories about when we all went to our first ISDN meeting. I said, "Well, I went to one 25 years ago," and he said, "27," so he had me—that kind of thing. ISDN is one of those things that still may happen, but in the meantime we have another revolution coming along beyond the basic digitization of the network—we have the packetizing of communication and ATM and the Internet.

Internet began growing in the 1970s, and now we think of it as exploding. Between Internet and ATM something is happening out there that is doing away with our fundamental concept for wired communications. First we did away with analog, and then we had streams of bits, but now we are doing away with the idea of a connection itself. Instead of a circuit with a continuous channel connecting sender and receiver, we have packets floating around disjointedly in the network, shuttling between switching nodes as they seek their separate destinations. This packetization transforms the notion of communication in ways that I don't think we have really come to grips with yet.

Where we stand in 1993 is this. All interexchange transmission is digital and optical. So is undersea transmission, which is currently the strongest traffic growth area: between nations we have increasing digital capability and much cheaper prices. The majority of the switches in the network are now digital—both the local and the tandem switches. But there is a very important point here. In these digital switches a voice channel is equated with a 64-kilobits-per-second stream. It is not as if there were an infinite reservoir to do multimedia switching and high-bandwidth applications, because the channel equals 64 kilobits per second in these switches. They are not broadband switches in terms of either capacity or flexibility.

Another important conceptual revision that has occurred during the digital revolution has been one involving network intelligence. The ESS number five, AT&T's local switch, has a 10-million instructions per second (MIPS) processor as its central intelligence. That was a big processor at the time the switch was designed, but now the switch finds itself connected to 100-MIPS processors on many of its input lines. The bulk of intelligence has migrated to the periphery, and the balance of the intelligence has been seeping out of the network. I always think of Ross Perot's giant sucking noise or, as George Gilder wrote, the network as a centrifuge for intelligence. This has been a direct outgrowth of the personal computer (PC) revolution.

THE BOTTLENECK: LOCAL LOOP ACCESS

Let us turn now to loop access, where I have said that the bottleneck exists. Loop access is still analog, and it is expensive. The access network represents about 80 percent of the total investment in a network. That is where the crunch is, and it is also the hardest to change. There is a huge economic flywheel out there to change access, whereas the backbone, as we have seen, can be redone in a matter of a few years at moderate expense.

There are many things happening in the loop today, but I see no silver bullet here. People always say there ought to be some invention that is going to make it cheap to get from the home into the network, but the problem is that in the loop there is no sharing of cost among subscribers. In the end you are on your own, and there is no magic invention that makes this individual access cheap.

Today everybody is trying to bring broadband access into the home, and everybody is motivated to do this. The telephone companies want a new source of growth revenue, because their present business is not considered a good one for the future. The growth rate is very small and it is regulated, and so they see the opportunity in broadband services, in video services, and in information services, and they are naturally attracted to these potential businesses. On the other hand, the cable television companies are coming from a place where they want to get into information services and into telephony services. So from the perspective of the telephone companies, not only is the conventional telephone business seen as unattractive, but there is also competition coming into it, which makes it even less attractive.

There are many alternatives for putting broadband service into the home. There are many different architectures and a number of possible media. Broadband can be carried into the home by optical fiber, by coaxial cable, by wireless, and even by the copper wire pairs that are currently used for voice telephony.

The possibilities for putting fiber into homes are really a matter of economics. The different architectural configurations differ mainly in what parts of the distribution network are shared by how many people. There is fiber to the home, fiber to the curb, fiber to the pedestal, and fiber to the whatever! The fact is that when we do economic studies of all these, they don't seem to be all that different. It costs about $1,100 to put a plain old telephone service (POTS) line into a home and about $500 more to add broadband access to that POTS line. You can study the component costs of all these different architectures, but it just does not seem to make a lot of difference. It is going to be expensive on some scale to wire the country with optical fiber.

The current estimate is that it would be on the order of $25 billion to wire the United States. This is incremental spending over the next 15 years to add optical fiber broadband. Moreover, we are probably going to do that not only once, but twice, or maybe even three times. I picture standing on the roof of my house and watching them all come at me. Now we hear that even the electric power utilities are thinking of wiring the country with fiber. It is a curious thing that we are going to pay for this several times, but it is called competition, and the belief is that competition will serve the consumer better than would a single regulated utility.

Because the investment in fiber is so large, it will take years to wire the country. Figure 1 shows the penetration of fiber in offices and the feeder and distribution plants by year, and the corresponding penetration of copper. You can see the kind of curves you get here and the range of possible times that people foresee for getting fiber out to the home. If you look at about a 50 percent penetration rate, it is somewhere between the year, say, 2001 and 2010. This longish interval seems inconsistent with what we would like to have for the information revolution, but that is what is happening right now with the economic cycle running its normal course.

The telephone company is not the only one that wants to bring fiber to your home. The cable people have the same idea, and Figure 2 shows the kind of architecture they have. They have a head end with a satellite dish and a video-on-demand player, or more generally a multimedia server. They have fiber in the feeder portion of their plant, and they have plans for personal communication system (PCS) ports to collect wireless signals. They have their own broadband architecture that will be in place, and, of course, we see all the business alliances that are going on between these companies right now. We read about these events in the paper every morning.

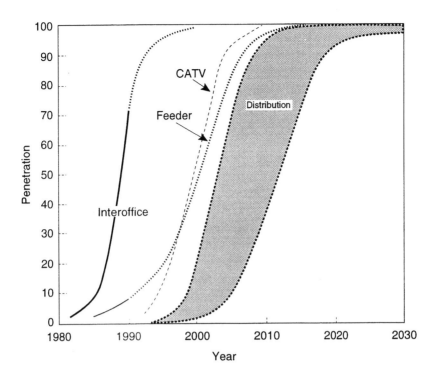

FIGURE 1 Evolution of fiber in the local exchange network. Broad solid line, fiber; thin solid line, copper. CATV, cable television. Courtesy of Bellcore.

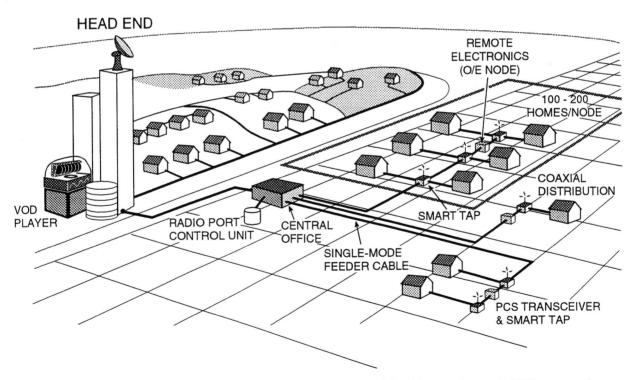

FIGURE 2 Fiber/coaxial bus cable broadband network. VOD, video on demand; PCS, personal communication system. Courtesy of Bellcore.

Everybody thinks that the money will be in the provision of content and that the actual distribution will be a commodity that will be relatively uninteresting.

If you look at the cable companies versus the telephone companies, cable has the advantage of lower labor costs and simpler service and operations. Their plant is a great deal simpler than that of the telephone companies. They have wider bandwidth with their current system and lower upgrade costs. They are also used to throwing away their plant and renewing it periodically. Disadvantages of cable franchises have more to do with financial situations, spending, and capital availability.

THE PACKETIZING OF COMMUNICATIONS: ATM

Asynchronous transfer mode (ATM) is a standard for the packetizing of communications, where all information will be carried in 48-byte packets with 5-byte headers. This fixed cell size is a compromise: it is too small for file transfers and too big for voice. It is a miraculous agreement that we do it this way and that, if everybody does it this way, we can build an inexpensive broadband infrastructure using these little one-size-fits-all cells. But it seems to be taking hold, and both the computer and communications industries are avidly designing and building the new packet infrastructure.

ATM has an ingenious concept called the virtual channel that predesignates the flow for a particular logical stream of packets. You can make a connection and say that all the following cells on this channel have to take this particular path, so that you can send continuous signals like speech and video. ATM also integrates multimedia and is not dependent on medium and speed.

Since ATM local area networks (LANs) are wired in tree-like "stars" with centralized switching, their capacity scales with bandwidth and user population, unlike the bus-structured LANs that we have today. Moreover, ATM integrates multimedia, is an international standard, and offers an open-ended growth path in the sense that you can upgrade the speed of the system while conforming to the standard. I think it is a beautiful idea, as an integrating influence and as a transforming influence in telecommunications.

THE NEW INFRASTRUCTURE

Now let us turn to the other force—Internet. The Internet will be the focus of a lot of our discussions, because when people talk about the NII, the consensus is growing that the Internet is a model of what that infrastructure might be. The Internet is doubling in size every year. But whether it can continue to grow, of course, is an issue that remains to be discussed, and there is much that can be debated about that.

To describe the Internet, let me personalize my own connection. I am on a Bellcore corporate LAN, and my Internet address is rlucky@bellcore.com. At Bellcore we have a T-1 connection, 1.5 megabits per second, into a midlevel company, JVNC-net at Princeton. Typically, these midlevels are small companies, making little profit. They might work out of a computer center at a university and then graduate to an independent location, but basically it is that kind of affair. They start with a group of graduate students, buy some routers, and connect to the national backbone that has been subsidized by the National Science Foundation.

THE USER VIEW OF INTERNET ECONOMICS

As for the user view of the Internet, here is what I see. I pay about $35,000 for routers and for the support of them in my computing environment at Bellcore. But people say that we have to have this computer environment anyway, so let's not count that against Internet. That is an interesting issue actually. I don't think we should get hung up on any government subsidies relating to Internet operations, because if we removed all the subsidies it really would not change my cost for the Internet all that much.

All we have to do is take the LANs we have right now, the LAN network that interconnects them locally, and add a router (actually a couple of routers) and then lease a T-1 line for $7,000 a year from New Jersey Bell. That puts us into JVNC-net. We pay $37,000 a year to JVNC-net for access to the national backbone network. For this amount of expenditure about 3,000 people have access to Internet at our company—all they can use, all they want.

A friend recently gave me a model for Internet pricing based on pricing at restaurants. One model is the a la carte model; that is, you pay for every dish you get. The second model is the all-you-can-eat model, where you pay a fixed price and eat all you want. But there is a third model that seems more applicable to Internet. If you ask your kids about restaurant prices, they will say that it does not matter—parents pay. Perhaps this is more like Internet! The pricing to the user of Internet presents a baffling dilemma that I cannot untangle. It looks like it is almost free.

Internet is a new kind of model for communications, new in the sense that while obviously it has been around for a while, it is very different from the telecommunications infrastructure. If I look at it, what I see inside Internet is basically nothing! All the complexity has been pushed to the outside of the network. The traditional telecommunications approach has big switches, lots of people, equipment, and software, taking care of the inside of the network. In the Internet model, you do not go through the big switches—it just has routers. Cisco, Wellfleet, and others have prospered with this new business. Typically, a router might cost about $50,000. This is very different from the cost of a large telecommunications switch, although they are not truly comparable in function.

With a central network chiefly provisioned by routers, all the complexities push to the outside. Support people in local areas provide such services as address resolution and directory maintenance at what we will consider to be the periphery of the network. The network is just a fast packet-routing fabric that switches you to whatever you need.

THE CONTRAST IN PHILOSOPHY BETWEEN
THE INTERNET AND TELECOMMUNICATIONS

The contrasting approaches and philosophies are as follows. The telecommunications network was designed to be interoperable for voice. The basic unit is a voice channel. It is interoperable anywhere in the world. You can connect any voice channel with any other voice channel anywhere. Intelligence is in the network, and the provisioning of the network is constrained by regulatory agencies and policies.

Internet, by contrast, is interoperable for data. The basic unit is the Internet protocol (IP) packet. That is in its own way analogous to the voice channel; it is the meeting place. If you want to exchange messages on Internet, you meet at the IP level. In Internet the intelligence is at the periphery, and there have not yet been any significant regulatory constraints that have governed the building and operation of the network.

Pricing is a very important issue right now in Internet. The telecommunications approach to pricing is usage based, with distance, time, and bandwidth parameters. Additionally, there are settlements between companies when a signal traverses several different domains. The argument that the telecommunications people use is that pricing regulates fairness of use. They feel that without some usage-based pricing, people will not use the network services fairly.

In Internet the policy has been to have a fixed connection charge, which depends on pipe size and almost nothing else. There are no settlements between companies, and there is no usage-based charge. The Internet fanatics would say that billing costs more than it is worth and that people use the network fairly anyway. That has worked so far, but it remains to be seen, as the Internet grows up—and especially as multimedia go on it—whether that simple pricing philosophy will continue to function properly.

Before I conclude I would like to introduce an issue that bothers me as a member of the telecommunications community. There was an article in the *Wall Street Journal* that said people are now bypassing the telecommunications network by sending faxes over the Internet. Here is the situation: if you send a fax over the Internet, it really appears to be free. You go to the telephone on your desk, and if you send the same fax, it costs you a couple of dollars. Why is this? This paradox sums up the idea that Internet really is almost a free good. Is the difference in price for the fax because Internet has found a cheaper way to build a telecommunications network? Or is it because Internet is a glitch on the regulatory system, that it is cream skimming and that it evades all the regulation and all the shared costs of the network? I can tell you honestly that I don't understand this question. I have had economists and engineers digging down, and I keep trying to think there will be bedrock down there somewhere. I cannot find it. The problem is anytime an engineer like me tries to understand pricing, I have the idea that it ought to have something to do with cost.

Unfortunately, that is not really the case. The biggest single difference we were able to find between the cost of a fax on Internet and on the telephone network was that in the cost of the business telephone there is a large subsidy for residential service built into the tariff. In traditional telecommunications the relationship between the cost of providing a service and the pricing for that service is sometimes seemingly arbitrary. Furthermore, the actual cost of provision may involve a number of rather arbitrary assumptions and in any case is only weakly dependent on technology. At base, telecommunications is a service business.

The arresting fact about Internet is its growth rate of 100 percent. There are approximately 2 million host computers connected to it. It is about 51 percent commercial usage. Moreover, it violates the rules of life—someone must be in charge, someone must pay. Very troublesome. Yet it is a worldwide testbed for distributed information services, and it is forming the prototype information infrastructure of our time.

MULTIMEDIA AND THE FUTURE OF THE INTERNET

As we look to how the Internet might have to evolve in the near future, we see that the IP protocol that has grown up with Internet is going to have to change in some small but fundamental way. As it stands now, the basic unit is the connectionless data packet, and that does not really support video and voice on Internet in spite of the fact that, as we all know, those services are being sent experimentally on Internet right now. However, they only work when there is very little traffic on the network. If the traffic grows, voice and video cannot really be sent on Internet because it presently uses statistical multiplexing of packets, so they don't arrive regularly spaced at the edges of the network. So the future IP protocol has to have some kind of flow mechanism put into it to assure regularity of arrival of certain packets. Furthermore, the new protocol must support more addresses than are presently possible. But these problems are fixable. Security is also often cited as a problem, but that too can be fixed.

Communications people are betting that the future traffic will include a lot of multimedia. That means a great deal of additional capacity will be required, as well as control mechanisms appropriate to the new media. ATM is the answer being propagated by telecommunications companies, but Internet people say ATM does not really matter, because that is a lower-level protocol to TCP/IP.

Meanwhile, the computer industry is building new LANs that take ATM directly to the desktop. It now seems that we are headed to a situation where we will have ATM in the local LANs, but to be internetworked that ATM will be converted to TCP/IP to enter the routers that serve as the gateways to the network beyond. From there, however, it might again be converted to ATM in order to traverse the highways of the network. Similarly, on the other half of its voyage, it would again be converted back to IP and then to ATM to reach the distant desktop. That seems like a lot of conversions, but they will be easy in the immediate future and may well be the price of interoperability.

The prize in the future is control of multimedia—whether telecommunications controls multimedia through ATM or whether Internet is able to gear itself up, change its protocol, and capture this new kind of traffic. Will people use the Internet for voice five years from now, and if so, will that take any significant traffic away from the network? Those are interesting questions that I think we can debate.

The Search for the Holy Paradigm: Regulating the Information Infrastructure in the 21st Century

Charles M. Firestone

The thrust of this paper is that the electronic information infrastructure, which I define as the production, electronic distribution, and reception of information, has undergone two major regulatory paradigms and is ready for a third. The first is the rather legalistic reaction to monopoly and oligopoly and is manifest in the period of antitrust and regulatory enforcement under the Communications Act, culminating in the late 1960s to early 1970s. The second is the competitive period with its commensurate deregulatory paradigm, which has been present in communications regulation since the late 1970s. These overlapping stages find expression in the regulation of many different aspects of the information infrastructure.

My thesis is that these prior models are inadequate, by themselves, for the complexities of the 1990s and beyond. While I do not specify a definitive new paradigm, I suggest that we might look to scientific or ecological models in beginning to sort out ways for the government to interact, in a dynamic democratic process, with the coevolving technologies, applications, and regulatory environments of the communications and information world: the international information environment within which the national information infrastructure will wind. In that sense, I call for a vision of "sustainable democracy," wherein technological developments preserve and enhance present and future democratic institutions and values.

DEFINITION OF THE INFORMATION INFRASTRUCTURE

The information infrastructure can be broadly or narrowly defined. For the purposes of this paper, I take a broader outlook that corresponds to the nature of the communications process, namely, the production, distribution, and reception of information in electronic form. Thus, while I will allude to print and newspaper, the focus will be on the following elements of electronic communications:

• Production of information in film, video, audio, text, or digital formats;

• Distribution media, most notably telephony, broadcasting, cable, other uses of electronic media, and, for our purposes here, storage media such as computer disks, compact disks, and video and audio tape; and

• Reception processes and technologies such as customer premises equipment in telephony, videocassettes and television equipment, satellite dishes, cable television equipment, and computers.

Others might prefer to distinguish only between software and hardware, between content and conduit, or between information and communications.[1] While such an organization is a plausible one, I favor the tripartite approach because it places attention on an increasingly important but often overlooked area of regulation, that of reception. As First Amendment cases move toward greater editorial autonomy by the creators, producers, and even the distributors of information, attention will have to be focused on reception for filtering and literacy concerns (Aufderheid, 1993). I address this in greater detail below.

REGULATION OF THE INFRASTRUCTURE

Regulation of the information infrastructure can also be usefully analyzed from this three-pronged approach. The law of production is associated with information law, with intellectual property concerns, First Amendment cases, and issues of content. Governmental regulatory policy in this area has also been affected by the expenditure of billions of dollars in governmental and public investment in programming for public broadcasting, research and development grants, creation of libraries, and software generation by the government.

Regulation of telecommunications has centered, however, on the distribution media. For many years this took the form of categorizing the form of media—telephony, broadcasting, or print—and placing that medium in the proper regulatory mold, namely, common carrier, trusteeship, or private. The main function of government was to make the electronic media usable, to allocate and assign uses and users to the particular media, and then to regulate the licensees in the public interest.[2] The major intention of this paper is to describe the past, present, and future paradigms of such regulation. At the same time, I hope to point out other levers of governmental policy, including investment in technology, or other nonregulatory influences.

Regulation of the receiver or reception process is relatively unusual, but important nevertheless. One familiar area was the early regulation of cable television, which was a technology intended to enhance reception of over-the-air television stations but was regulated to ensure that, at first, it did not undermine the economics of the broadcast television system.[3] Most significant in recent forms of regulation of reception is the area of privacy, which, among other rights, protects the receiver to be free from intrusion at the reception end of a communication.[4]

Goals of Regulation

In each case the aim of regulation was to effectuate or bring about a set of goals for society. I find it useful to categorize these goals into five basic democratic values, as follows.[5]

Liberty

Foremost among the Bill of Rights is the First Amendment guarantee against abridgment of freedom of expression (speech, press, association, petition, and religion). Beyond this trump card of protection, certain regulations have sought to promote First Amendment values of editorial autonomy,[6] diversity of information sources,[7] and access to the channels of public communication.[8] A long-standing debate centers on the proper legal attention that should be paid to the underlying values of the First Amendment, as opposed to the bare dictates of the amendment, that is, is it a sword as well as a shield?

A second liberty, implicit in the Bill of Rights but explicit in state constitutions and in legislation, is the right to privacy, individual dignity, and autonomy. At the federal level this has been expressed in the form of sector-by-sector legislative codes; at the state level privacy is protected in common law and statutes.[9]

A third liberty, which is not often expressed as such, is that of ownership of property. In the world of communications this usually takes the form of a bundle of intangible or intellectual property rights that attach to original works of authorship or invention by the Constitution's copyright and patent clause. Traditionally, individuals have been licensed as trustees but not owners of the electromagnetic spectrum. However, this approach could change with the auctioning of spectrum. Furthermore, the application of intellectual property concepts to new forms of information has raised many new and vexing problems for future regulators.

Equity

A countervailing value to liberty is that of equality. In the field of communications this takes the form, most prominently, of universal service. That is, the Communications Act gives, as one of its seminal purposes, to make available to "all the people" a worldwide communications system.[10] Today, the issue of universal service is foremost in the minds of regulators and legislators, who see the need for an evolving definition of which communications and information services should be available to all. Meanwhile, traditional schemes for assuring universal service are dissipating.

Another issue that comes within the equity value is that of information equity—is the gap between rich and poor extending to the information haves and have-nots? Are the technologies that have such potential to bridge the gap between rich and poor instead going to widen it? Governments at all levels have established libraries to provide all citizens with access to a broad array of information resources. The obvious question for the future is the level and extent that information resources will be made available in nonprint formats, and through electronic connections to the home, rather than by extensive collections at libraries.[11]

Community

The mass media, long attacked for monolithic mediocrity, is also seen as a key cohesive force in a community. The Federal Communications Commission (FCC) has included, as a basic tenet, the concept of localism as a touchstone for licensing and regulating broadcasting stations.[12] This includes a preference for locals obtaining the licenses initially and for stations, once licensed, to serve the local needs and interests of the service area.[13]

Similarly, franchising authorities often grant cable television franchises to locals, who are thought to be more attuned to the needs of the local community. Whoever receives the franchises, furthermore, usually has requirements to include public, educational, and governmental access channels.[14] At one time, the FCC required cable operators to include local origination cablecasting channels on their systems.[15] The access channels, as well as some public broadcasting and other cable channels, might be viewed as electronic commons or public space for the local community.

At the same time, on a national scale, the broadcast media serve as one of the country's most significant and binding cultural forces. The broadcast media provide common experiences and a larger, more cohesive national community for a country that is increasingly multicultural.

A third manifestation of community values is the advent of public broadcasting—governmental and public financial support of noncommercial fare. Some of that programming is aimed at better serving local communities. This aspect of "community" raises a broader issue for the future: What is the role of government financing of programming, information resources, and distribution facilities?[16]

Finally, the whole concept of community is changing in the electronic world of networks and targeted communications. New electronic communities are being formed around the interests of individuals rather than geographic boundaries. This, again, raises profound questions about the impact of modern communications technologies on the concept of community.

Efficiency

The Communications Act specifically refers to the need for an "efficient" communications system by wire and radio.[17] Generally, economic regulation values efficiency as its ultimate goal and fairness secondarily. Much of the communications regulatory scheme is premised on making the infrastructure and the pricing system efficient. Title II common carrier regulation seeks to assure that telecommunications facilities serve the entire country at rates that are fair, affordable, and comparable to those that would result from competition.[18]

In the wireless realm, governmental spectrum allocation is premised on the tenet that, if unregulated, spectrum users would injuriously and therefore inefficiently interfere with each other.[19] More broadly, the government considers how communications and information sectors can maximize the citizenry's economic welfare, within the country and on a more global scale. Recently, the United States has been concerned with its competitive position in the world economy. The promotion of an efficient communications and information infrastructure supports that concern and interest.

Participatory Access

The communications system needs to be not only efficient but also workable and accessible to ordinary citizens. Accessibility is an undercurrent in many elements of communications regulation, from the creation of free over-the-air broadcasting stations, to access channels on cable, to common carrier regulation of telephony and its progeny. Indeed, access could be an organizing principle for regulation, in terms of facilities, costs, and, as will be explained later, literacy (Reuben-Cooke, 1993).

Finally, the regulatory scheme seeks to enable citizens to participate in the process on a fair and equal basis. The concepts of due process, sunshine for governmental agencies, and appellate review are all designed to facilitate citizens' fair access to, and participation in, the process. By using new communications technologies, these goals will likely be enhanced.

Whatever regulatory scheme is employed, it will have to consider and balance the above basic values and goals to fashion a system of greatest benefit to the public. The following sections will describe how these goals were addressed in the regulatory paradigms of yesterday and today, along with how they might be considered and treated in the future.

REGULATORY PARADIGMS

At each level of the communications process—production, distribution, and reception—there have been two paradigmatic stages of regulation, one of scarcity and one of apparent abundance and competition. At the scarcity stage, usually obtained after an initial period of skirmishing among pioneers for position, the regulation has taken the general form of governmental intervention in order to promote the broader public interest.[20] At the abundant or competitive stage, which overlaps the earlier stage, there has been a reversal, a deregulation to promote greater efficiency (Kellogg et al., 1992). In each case the paradigm is a regulatory religion, at times demanding faith on the part of the believers. My thesis is that neither of the past regulatory schemes is sufficient alone to address the evolving complexities of the new communications infrastructure and that a new regulatory paradigm, a new religion, will be needed for the global and digital convergence of the telecommunications future.

Stage 1: Scarcity

The overriding characteristic of the first stage, "scarcity," is that in the various levels and industries associated with the information infrastructure there were large centralized, monolithic, top-down, command-and-control-type enterprises, relatively passive receivers, the formation of monopolies or oligopolies, often of a vertical (end-to-end) nature, and a certain potential for anticompetitive practices. These could be symbolized by the early stages of the major movie studios, AT&T, the broadcast networks, and IBM.

In most if not all cases the tendency of the law was for the government to regulate the monopoly industry in the public interest, either through application of antitrust laws or the regulatory apparatus of the Communications Act. Usually, this resulted in the removal or regulation of bottlenecks, to try to approximate the efficiencies of competition where none existed. The Communications Act also sought to foster the values of equality (in the form of universal service) and community (in the form of localism) under the general regulatory standard of the "public interest, convenience, and necessity."

Production

The production of information has been the most widespread, abundant, and competitive of all of the elements of the infrastructure. In almost each case the production industries have not been subject to regulation, for First Amendment and other historical reasons. Rather, they are subject to normal commercial, antitrust, intellectual property, and First Amendment laws. Nevertheless, where economic oligopolies arose at earlier stages of the film, telephone technology, and computer equipment industries, the antitrust laws were used to address the alleged monopolization. Although the antitrust cases had only mixed success, that was the accepted method for governmental activity in this area. In the area of intellectual property, the law imposed a form of regulation in the imposition of fair use and compulsory licensing prescriptions. In the lone industry subject to federal agency regulation—broadcast programming—the FCC imposed both antitrust-like and content-related remedies.

Motion Pictures. In the 1930s and 1940s the major motion picture studios found that vertical integration could enhance their ability to control their products from inception to exhibition. (Some studios, such as Warner Brothers, were formed by theater chains.) The Justice Department, however, alleged that the studios were monopolizing the industry by both their structure and certain anticompetitive practices.

In the *Paramount* consent decree of 1948,[21] the major, vertically integrated motion picture production studios and distribution organizations of the time agreed not to monopolize the production, distribution, and exhibition sectors of the motion picture industry and to divest their theater chains. Thus, a bottleneck at one level (e.g., major movies or control of theaters) could not be used to create scarcities and affect competition at another. This case is instructive to our inquiry, then, for its principle: to assure competition at each level of the communications process. Today, the analogy would require competition (or openness) at the production, distribution, and reception levels of all communications.

Telephone Equipment.[22] Shortly after the Justice Department litigated the *Paramount* case, it brought an action against AT&T and its Western Electric subsidiary for using its patents to monopolize the telephone equipment business.[23] In an antitrust consent decree—the original "Final Judgment"—AT&T agreed to segregate the company into strictly regulated businesses, forsaking competitive businesses.[24] Western Electric's equipment would henceforth be sold only to Bell companies. Significantly, it also required that the patents be licensed to other manufacturers, a form of compulsory licensing.[25]

Broadcast Programming. The broadcasting networks were regulated by the FCC and thus were subject to a broader "public interest" standard. During the 1930s and into the 1940s, NBC controlled both a Red and a Blue network, with only modest competition from CBS. The parent network imposed a set of requirements on its affiliates to take what was offered. After a long inquiry, the FCC (1) ordered NBC to divest one network (thereby creating ABC), (2) restricted certain abusive network practices that took discretion away from local affiliates, and (3) prohibited local stations from affiliating with more than one network.[26]

Broadcast Content Controls. Content controls were imposed on broadcasters from the inception of the Radio Act of 1927 and the Communications Act of 1934. These included the requirement that a station afford equal opportunities to legally qualified candidates for public office to "use" the facilities[27] and restrictions against certain illegal activities such as obscenity.[28] After first prohibiting editorializing,[29] the FCC imposed a Fairness Doctrine, which encouraged stations to air controversial issues of public importance and to present contrasting viewpoints from responsible spokespeople on such issues.[30] Furthermore, at the height of content regulation, the late 1960s and early 1970s, the FCC's regulations extended to changes of formats,[31] to programming designed to respond to the needs and interests of the local community,[32] and programs on public affairs.[33]

Copyright. The most common form of "regulation" of the production of programming and other information is through governmentally protected ownership rights, in the form of intellectual property laws. In the field of communications the most prevalent legal issues are in the area of copyright, which relates to original works of authorship. Here, the law provides one with a monopoly over the use and distribution of the work for a limited term.[34]

But even in the area of intangible ownership, as in real property law, there are limits to the rights of the owner. Thus, the public may have limited rights to use or copy a work for public interest reasons,[35] just as it has rights of access to certain properties (e.g., private beachfronts in California) for public purposes. This area of "fair use" is vague and complicated, but even in the pure ownership of intellectual property there is a certain amount of regulation of a monopoly for public benefit.

Similarly, there are certain situations where the copyright owner is entitled to compensation for use of the work but cannot restrict to whom and where the use is extended. Thus, "compulsory licensing" has been extended to cable operators and jukebox owners in order for them to access certain works that are otherwise too cumbersome to license on an ad hoc basis.[36] (There are other instances of cumbersome licensing where compulsory licensing is not extended but, rather, private licensing agencies are employed. The compulsory licensing provisions of the Copyright Act are mainly the result of political compromises.)

Computer Manufacturing. In the second and third quarters of this century, IBM dominated the field of computer manufacturing. The IBM mainframe epitomizes the centralized, top-down, monolithic enterprise of the scarcity stage of regulation. The Justice Department, however, was unsuccessful in its antitrust suit against IBM and dropped the suit on the same day that AT&T agreed to the Modified Final Judgment.[37] Nevertheless there is a history of private antitrust suits that did affect, and in a sense "regulate," the computer industry.[38] More significant, however, was the research and development work in various government agencies, particularly the Defense Department, that led to many of the advances in computer hardware and software.

Distribution Media

In contrast to the production entities and businesses, the electronic distribution media during the 1920s through the 1970s were characterized by direct regulatory control of centralized gate-keepers under the public interest standard. Here the concepts of scarcity had two distinct meanings and regulatory approaches: common carriage, which was applied to telephony, and public trustee-ship, which was applied to broadcasting. These concepts are explained below.

Telephony. In the field of telephony the building of facilities and the operation of a telephone network were thought to have natural monopoly characteristics—large barriers to entry due to high initial capital costs and increasing economies of scale. From the time of Theodore Vail's 1909 use of the words "universal service" to explain his plan for exchanging governmental protection and regulation for the extension of service to all at fair rates,[39] the Bell system's end-to-end monopoly had social equity implications. Regulation of that monopoly as common carriers under a broad public interest standard would maintain an efficient, fair, reliable, accessible, and affordable telephone system for the country.

Under the Communications Act, common carriers have been required to apply for permission to enter or exit a market.[40] They may not discriminate among their customers; they must provide access to all at reasonable rates, which are reviewed by the government; and they must file tariffs of their rates and practices.[41] States generally paralleled this approach for intrastate communications.[42] So long as there was a monopoly carrier, the rates could be averaged among all the customers of a given area so that high-cost areas could be subsidized within the rate system by ratepayers from high-volume areas. This regulatory deal, which promised stability, growth, universal service, and restraint against competitors, reached its zenith in the late 1960s and early 1970s.[43]

Satellites. The use of satellites for communications common carriage derived from Arthur Clarke's model of covering the world by three geostationary orbiting satellites (Clarke, 1945).[44] The United States instigated the founding of INTELSAT, an international consortium of countries to foster international communications. The INTELSAT agreement contemplates a monopoly carrier to handle the domestic traffic to and from the INTELSAT satellites.[45] This vision of a regulated monopoly was perfectly consistent with the prevailing paradigm of the time. It followed, as well, the tremendous federal government investment in a space communications system. Thus, during the 1960s and early 1970s, COMSAT was the American monopoly satellite space segment provider.

Broadcasting. In the use of the electromagnetic spectrum it was long perceived as necessary to have the government allocate uses and users in order to avoid destructive interference on each frequency. For example, in the mid-1920s the broadcasters themselves called for federal regulation of the AM airwaves (Barnouw, 1975). Congress responded with the Radio Act of 1927 and the successor Communications Act of 1934, regulatory schemes whereby access to the airwaves was limited by law. Only those relative few who were licensed would be allowed to broadcast. These licensees of "scarce" frequencies, however, would be deemed as *trustees* for the public and therefore regulated according to the "public interest, convenience, and necessity."[46]

This regulatory scheme grew through the 1960s to incorporate various requirements on each broadcaster to serve its local community. It required stations to provide access to candidates for federal elective office;[47] to provide equal opportunities to opposing candidates for the use of the station's facilities ("equal time");[48] to provide its lowest unit charge for political advertising;[49] to air contrasting viewpoints on controversial issues of public importance;[50] to ascertain the local needs and interests and to design programming aimed at meeting those ascertained needs,[51] including those of minority audiences;[52] to air news and public affairs programs;[53] to have some prime time available for nonnetwork programs;[54] to refrain from airing obscenity,[55] indecent language,[56] lotteries,[57] fraudulent programming,[58] or too many commercials;[59] and to concern themselves with many other regulations considered to be in the public interest.

These requirements were far more stringent than those applicable to newspapers, under the theory that as licensees of scarce radio frequencies, broadcasters must operate under a regime where the public's interest is paramount.[60] In the newspaper business, on the other hand, the editorial discretion of the publisher is the primary legal consideration.[61] While the scarcity theory has been seriously questioned by many, it is still the applicable law today.

Cable Television. As cable television grew from the hilltops in rural areas to the wiring of towns and cities, local jurisdictions were called upon to franchise cable operators' uses of the streets and rights of way. These franchises were often exclusive, whether de jure or de facto. Once an operator obtained a franchise, it was only rarely that a locality granted another franchise for the same area, even if competitors knocked on the door. Again, like the monopoly distribution media described above, franchising authorities exerted certain regulatory controls over the franchisees.[62] At first these included rate regulation; later, and particularly after rate regulation was preempted at the federal level,[63] franchising authorities imposed other public interest obligations on the cable operators, which at times took the form of payments to the city for noncommunications-related activities.

Private Carriage. A third type of media entity gained prominence during the 1970s—private users of the spectrum and, eventually, private wire-based facilities providers. While spectrum users had to obtain licenses and were still subject to public interest regulation, the burdens were minimal and technical in nature. Primarily, these entities were treated under the *private* model of regulation, left to use the medium for the purpose for which it was granted (i.e., a trucking company could obtain frequencies for two-way radiotelephone communications but could not use them for nonbusiness purposes and definitely could not "broadcast" on them).[64]

Storage Media. Records, tapes, and disks, like print media, have been subject to the "private" model of regulation—application of antitrust and other laws generally applicable to all businesses.[65] While the government does not directly regulate, it can potentially impact the production and distribution of these media through large or strategic purchases of selected titles. Thus far, however, government purchases of these media are mainly through normal library procedures.[66]

Reception

At the scarcity stage, the law of reception was primarily concerned with integrity of the respective systems of communication. With respect to telephony, AT&T retained through its tariffs and its regulatory deal the ability to restrict any foreign attachment to its facilities. This went so far as excluding plastic book covers for the yellow pages directory and cups on handsets to prevent others from hearing a conversation. (For the latter, called a "hush-a-phone," it was possible to obtain appellate court reversal of the FCC's enforcement of AT&T's tariffs under the determination that if something is beneficial and not publicly detrimental, it could be added.)[67]

A second major battleground during the 1950s and 1960s was the use of community cable television (CATV) antennas to improve reception of nearby television stations. The FCC flip-flopped on its attitude toward CATV but eventually held that it needed to be regulated as ancillary to the commission's power over broadcasting.[68] In 1968 the Supreme Court held that cable systems were enhancers of the reception mechanism for television, rather than rebroadcasters, for the purposes of the copyright laws.[69] The commission then proceeded to regulate these reception enhancers so as to prevent importation of certain distant television stations or to unduly fragment the audiences of local stations servicing a particular area.[70]

Other regulatory decisions affected reception, such as the FCC's decisions selecting a particular system for color television[71] and Congress's requirement that television sets include UHF channels.[72] The color television episode was an example of the government setting a standard: some in the industry win, some lose, but the determination is made by the government.

Summary of Stage 1

In sum, the period of the 1920s to the 1970s was one of regulation in the telecommunications business according to the touchstone of the Communications Act—the "public interest, convenience, and necessity." The major question, in each case, was, What was the "public interest"? This standard, according to the Supreme Court, was a supple instrument for guiding the industries in periods of rapid technological change.[73] Nevertheless, it was admittedly vague. Thus, the agenda was essentially set by the lawyers, who would argue as to how far this vague and lofty standard should or could extend before it exceeded the bounds of regulatory authority or the limits of the First Amendment.

In addition, the antitrust laws were used to restrict economic abuses and to promote access to bottlenecks. Through consent decrees, the Justice Department and courts served as a kind of regulator of the movie and telephone equipment industries. Analogous regulation also appeared in other contexts, such as the fair use doctrine in copyright.

Finally, state, local, and private parties enforced other applicable regulatory regimes. Governments had many levers to pull and buttons to push in shaping the form and substance of the communications infrastructure. These included direct regulation, governmental investments, research grants, low-interest loans, tax incentives, and even the bully pulpit.

Despite the logic and general agreement that the public interest was the paramount consideration, this general regulatory regime began to unravel in the 1970s (with certain prior indications of problems well before that). Scarcity was questioned both technologically and philosophically. A deregulatory mood swept the country, symbolized by the election of Jimmy Carter as President. Regulatory systems, such as trucking and airlines, as well as telecommunications, were seen as less efficient than they would be under competition. There was a sense that a new approach would resolve the country's regulatory interests better than a cumbersome, bureaucratic, and failing system. But most particularly, there were significant technological breakthroughs in microwave, broadband communications, and computerization that challenged the natural monopolies, oligopolies, and technological underpinnings of scarcity.

Stage 2: Abundance and Competition

In each of the affected industrial sectors under review, the key industrial elements became abundant, competitive, or both, creating the need for a new regulatory paradigm. Mostly, this process became manifest in the 1970s, strong in the 1980s, and perhaps excessive by the 1990s. I will look at each sector individually, pointing out, where appropriate, the parallels among industrial sectors.

Production

Compared to the distribution media, the production elements of the communications process were essentially competitive even during the "scarcity" stage. But by the 1970s, developments in these industries created new levels of competition and commensurate regulatory responses.

Motion Pictures. We begin with the motion picture industry. The *Paramount* decree was adopted just as television became a consumer product. With the steady increase in viewership of television, the film industry suffered box office declines for a while. But as the studios (1) sold movies to television as a new window for their products and (2) moved into the production of television programming, they saw a resurgence in their businesses (Owen and Wildman, 1992).[74] With more outlets (theater and television), there arose more production entities. Talent agencies, independent producers, and syndicators all added levels of production entities for both film and television. As new windows arose, with pay television, pay cable, and videocassette tapes, still more entities entered the businesses, many of which also produced film, television, and tape. The advent of increasing competition for production, characterized by the independent television and film producer, created an environment of competition. This in turn led to a reintegration at the edges of the new media. Paramount and MCA jointly owned the USA cable network. Other motion picture companies began to own other outlets, such as the Disney Channel. 20th Century Fox was heavily involved in film, television, and videotape production. CBS tried the film business for a while but was unsuccessful. Vestron, an early leader in the videotape business, also tried but failed in the film business.

The *Premiere* case[75] of the early 1980s was the final vestige of Stage 1 antitrust regulation. The major studios, buoyed by the trend toward reintegration and the governmental attitude of relaxed enforcement of antitrust laws, entered into an arrangement whereby the studios would provide their best movies to their jointly owned and operated Premiere pay cable network. The court determined that the studios were prima facie violative of the antitrust laws. The fledgling network was disbanded, but the film and television production entities have since become more and more integrated. The *Paramount* decree has become all but a dead letter, and economic regulation has become anathema in the film industry.

For whatever reason, the expanded world of television appears to be producing a diet of more violent and sexual programming. Public reaction to this has taken the form of calls for content controls to minimize violent programming. But even here the call for regulation has recognized a competitive paradigm in production rather than Stage 1-type content controls at the production levels. For violent programming the calls have been to rate programs, a measure aimed at enhancing consumer information instead of restricting output. Another possibility is to require a chip in the receiving apparatus so that the receiving home can edit objectionable programming. These measures, then, are aimed at enhancing competition and markets, consumer choice, and freedom of individuals to receive access to a diverse array of products.

Copyright. The production of information enhanced by the copyright laws has also seen greater competition and reduced regulation in recent times. Fair use applications of the copyright laws became less available after passage of the 1976 Copyright Act amendments, which set forth a set of guidelines for the doctrine and expanded the bundle of rights contained in an author's claim to ownership of his or her original works of expression.[76] This has resulted in greater use of contracts, a classic marketplace mechanism, as a means for determining the rights and responsibilities of owners and users of copyrightable material. In the field of cable television, one of the mainstays of compulsory license, cable's carriage of local television signals, was altered by new provisions requiring cable operators to obtain "retransmission consent" from the owner of those signals.[77]

Computers. The paradigm of competition also manifested itself in the unregulated computer industry. IBM faced competition not only from other manufacturers of mainframes and large top-down-type computers but also from clone manufacturers, Apple, and others. As computer technology became smaller and cheaper, moving to the minicomputer to the microcomputer, competition grew in all aspects and levels of the industry. Indeed, the industry is a metaphor for the whole era in telecommunications regulation: increasing competition, less regulation.

Distribution Media

The heavily regulated distribution media saw the most dramatic paradigm shift in economic structure and regulation.

Telephony. First, two sets of decisions beginning in the late 1960s signaled the decomposition of AT&T's monopoly. The decisions freeing the terminal equipment market from regulation are described below in the "Reception" section. The other decision was to allow competition to AT&T in the intercity, long-distance markets.[78] This determination was essentially inevitable as a result of the advent of microwave and satellite transmission technologies. Microwaves allowed signals to be transported cheaply by line-of-sight relays over large distances. The long lines of old became short waves. Costs declined rapidly, and innovative entrepreneurs such as William McGowan of MCI saw some openings.

Long Distance. After the FCC allowed private entities to use microwave frequencies above 890 MHz to build their own private transmission facilities,[79] AT&T filed tariffs setting discounts for large users. But these were inflexible tariffs, and MCI sought to customize transmission bandwidth for large customers over its own competitive facilities. In 1969 the FCC allowed MCI to build a line from Chicago to St. Louis.[80] Later, it determined that competitors need not show the need for a new service in order to obtain a license as a Specialized Microwave Common Carrier to compete with AT&T for long-distance business.[81] MCI eventually won the ability to offer switched long-distance service,[82] and others, such as Southern Pacific Railroad, followed with national long-distance networks of their own (which became Sprint). The flood of competition meant not only an open, competitive market for large users, who could pick and choose telecommunications providers and the types of service offerings they might want, but also competition and lower rates for the average subscriber—a sacrosanct market for the internal cross-subsidies contained in the regulatory regime of the scarcity stage.

Satellites. Domestic satellites, at the same period of the early 1970s, won an "Open Skies" decision by the FCC.[83] As long as they were legally, technically, and financially qualified, private entities could own and operate a communications satellite. Here, the government again moved to the competitive paradigm. The domestic satellite business would be a business, competitive with wireline services but free of many of the regulatory hamstrings inherent in the common carriage of old. The 22,400 feet to and from the satellite that each signal had to travel made distances on Earth insignificant. "Long distance" became a misnomer. The satellite carriers, who were already competitive in the sense of the Open Skies decision, also found increasing competition—first from resellers of satellite transponders, who were deregulated by the FCC,[84] and then from decisions allowing private competition to INTELSAT on the international stage.[85]

Local Loop. As the interexchange and terminal equipment parts of the telecommunications network became competitive and deregulated, the last frontier for competition was the heavily regulated local loop. Even here, competition has emerged in recent years (Entman, 1992). Two cellular franchises were granted for each market, one to local wireline carriers and one to competitors.[86] Now, after many years of building, selling, and trading, culminating in AT&T's purchase of the McCaw systems, the cellular business is not only competitive but also most formidable. Newer competitors are on the horizon, whether they come in the form of a third cellular license in each market, specialized mobile services, new Personal Communications Services (PCSs),[87] or other, even newer technologies.

Furthermore, the reduced costs of fiber optic transmission, and FCC orders requiring interconnection to local exchange facilities,[88] have also made alternative carriers at the local level an economically viable and burgeoning business. As cable television systems enter the local telecommunications business, with fiber backbones and broadband connections to 65 percent of the country, and as PCSs become a reality, these trends are only going to continue. Thus, even the heavily regulated local loop is beginning to enter the next stage of rapid competition and deregulation.

Television, Cable Television, and Multichannel Distribution. The competition phenomenon hit equally hard in the television delivery media. The number of local television stations steadily increased over the last four decades. This, coupled with the advent of cable television as a common delivery mechanism, has resulted in half the country's viewers having at least 20 channels available to them and over 40 percent with at least 30 channels.[89] With the increasing number of signals (over 80 national networks[90]), the FCC felt less obligated to regulate each one closely. Indeed, former FCC Chairman Mark Fowler called television a "toaster with pictures" and proceeded to "unregulate" as much as he could during the Reagan administration. This attitude received a healthy boost by the Cable Act of 1984, deregulating cable to a great measure.

Cable television's exclusive franchises have also been subject to several new competitors and challenges. Wireless cable (a compilation of multipoint microwave distribution channels) has finally entered the fray, and direct broadcasting satellites, long heralded but nonexistent until now, will at long last pose still another new threat to the multichannel delivery monopoly held by cable operators. Meanwhile, significant legal challenges to the local franchising monopoly are being waged both by upstart cable operators[91] and telephone companies.[92] Here the First Amendment, which was written as a prohibition against government infringements of freedom of communication, is being used to forge a wedge by industrial companies into a distribution business, with broad-ranging consequences.[93]

Reception

Reception devices and processes are becoming more important, for regulatory purposes, during the deregulatory era. The more responsibilities and choices that can be shifted to the consumer or reception end of the process, the less regulation is necessary or appropriate at the production, transmission, or distribution ends. Thus, the FCC and others are looking to creating greater choice and control at the user's discretion. This thirst to shift the regulatory burden to the consumer is being aided by advances in technology that allow for greater user control and targeting.

Terminal Equipment. The FCC broke AT&T's monopolization of terminal equipment on the Bell System in the *Carterfone* case in 1968.[94] There, relying on an obscure precedent from the 1950s, the commission allowed Carterfone to interconnect its device to the Bell network on the grounds that it would not be physically detrimental to the network and could enhance consumers' choices. So long as it was privately beneficial without being publicly detrimental, the foreign attachment could not be excluded.

This decision opened up the terminal equipment market and laid the groundwork for the commission's decision in *Computer Inquiry II*.[95] There the agency declared that customer premises equipment was to be competitive and therefore completely deregulated. The sole area of regulation was the commission's equipment registration rules, aimed at preventing harmful interference from such devices.[96] The ensuing terminal equipment business is now an open and competitive one, subject to only minimal regulations.

The AT&T divestiture proceeding in the mid-1980s prevented the divested Bell operating companies from entering into the equipment manufacturing business,[97] but this preclusion is now actively under review by Congress and the courts.

Privacy. The law of privacy has grown from a law review article in 1890 (Brandeis and Warren, 1890), to private tort law in the states, to state constitutional and statutory provisions,[98] to, eventually in the 1970s, a federal approach.[99] The federal laws, however, tend to follow industrial sectors, such as credit-reporting businesses,[100] financial records,[101] cable television,[102] and others.[103] The United States has thus far rejected the broad across-the-board regulatory approach that certain European countries have adopted (Reidenberg, 1992a,b). Typically, these federal laws will restrict the industrial company from maintaining certain files beyond a useful time period, will afford the subject individual with rights to ascertain what information is collected and maintained, to correct errors, and to have the information used only for the collected purposes, unless permission is granted for other uses.

Thus, in addition to their place in the constellation of human and (implicit) constitutional rights, privacy protections are one means of assuring individuals that their use of the information infrastructure will be secure. They have always involved balancing against other rights and interests and therefore a kind of regulation of the use of information.[104] While the timetable for regulation and deregulation is later than with the distribution media, there is discussion today of moving to a marketplace solution to some of the privacy issues in telecommunications.[105]

Reception of Indecent Programs and Messages. The law of indecency took shape in the late 1970s under a Supreme Court decision that established a nonscarcity basis for regulation. In *FCC v. Pacifica Foundation*,[106] the court held that the nature of the broadcast medium, intruding as it does into the home, allowed the federal government to impose "indecency"[107] restrictions on broadcast transmissions that are more stringent than those for print media.

But when prosecutors wanted to apply the same standards to cable broadcasts, the courts took a contrary, more relaxed approach. Because subscribers brought the signals into the home voluntarily, and because electronic lockboxes and other means of blocking signals were available to subscribers, the lower courts overturned regulatory attempts to block transmission of indecent programming over the newer technologies.[108]

Summary of Stage 2

In sum, spurred by technological innovations that expanded the opportunities to communicate over multiple paths, the second regulatory paradigm saw a perceived if not actual end of scarcity, an abundance of channels, frequencies, bandwidth, and competitors. The rationale for specialized economic regulation for the telecommunications industry became weaker and weaker, while the ideological winds of deregulation blew stronger and stronger. The result, beginning in the late 1970s and early 1980s, has been a period of strong deregulation in almost all telecommunications sectors, increasingly intense competition, and new economic relationships.

Nevertheless, the revolution of the newer competitive paradigm has not been absolute. Strong vestiges[109] of the first paradigm remain in the regulation of television (new requirements for educational children's television), reregulation of the cable television industry, continued maintenance of the Modified Final Judgment in restricting Bell operating companies from entering certain lines of business, extensive regulations imposed on those companies to assure that competitors can interconnect into the local exchange, protections of privacy interests, and calls for a renewed definition of the public interest in the field of communications regulation and policy.

Why can't the legislators get it straight and move all the way into the new paradigm? Some would fault special interest groups, industries each wanting their fair advantage, consumer groups wanting goods and services at unreasonably low rates. I would suggest something different. The thesis of this paper is that each paradigm properly has vestiges that remain during the next stage. But in this case the competitive paradigm is only a temporary stage, a kind of adolescence with certain traits that will continue throughout life.

The technological, political, economic, sociological, and scientific trends of the new age of complexity require a broader, more holistic approach to the subject matter. The second paradigm of competition met more carefully the goals of efficiency and liberty. The agenda, set by the economists, cleared away the underbrush of "externalities" that are not resolved by economic solutions. But in the 1990s this approach is simply too shallow for the issues and problems ahead. Most particularly, it does not resolve distributional problems adequately and slights the values of equality, community, and participation. As we perfect the solutions of the second paradigm of abundance and competition, we are already late for a new-age train moving out of the station carrying a variety of new trends and perspectives as its passengers.

THE NEW COMPLEXITY

In considering the applicability of the earlier regulatory schemes, one needs to explore the environment for which the regulations were designed. As I have detailed above, the lawyerly governmental intervention approach to regulation contained in the Communications Act of the New Deal was established in an era of scarcity of telecommunications resources and players. The public interest standard was designed to address goals of community, equity, and efficiency, while at the same time respecting liberties of communications.

As the environment changed, as communications channels became more plentiful, and as many new entrants came to offer communications services, the regulatory regime changed to a procompetitive, deregulatory milieu. This served primarily the goals of efficiency (as economists asked, What is competitive? What is efficient?), and liberty (freeing individuals to act as they chose, moving toward a laissez-faire approach).

Today, however, the environment is more rugged and complex than envisioned by the legal or economic theories of the prior regulatory ages. While remnants of the past paradigms will remain with us, and rightfully so, this new environment calls for a different approach to the government's interaction with the information infrastructure.

Technological Trends

Foremost among the drivers of change in this century are the tremendous technological changes. The technological trends that moved us from the first to the second paradigms were essentially those that made communications faster, shorter, cheaper, and better. These were the continual improvements on the processes of communication at the production, distribution, and reception stages. Transistors, microprocessors, microwaves, and even lasers and fiber optics all contributed to these advances.

The more current technological trends that are creating *digitization and convergence*, however, are creating a difference in kind more than of degree. By breaking voice, video, and data all into 1s and 0s, digital technologies are breaking down the ability to segment communications into neat regulatory cubbyholes. As Nicholas Negroponte suggests, eventually all transmissions will be bit radiation, which can be shaped at the reception end into the desired format (Negroponte, 1993).

This digitization, then, is creating a convergence of the production, media, and reception levels of the communications process. At the production stage, the movement to multimedia will further interrelate the voice, video, and data. At the distribution level, these digital bits can be packaged and sent through any pipes and decoded at the other end perfectly. And at the reception end, the converged component television-cable-computer will produce the product in the form desired by the user.

A second change in kind rather than degree is around the corner: the movement to *broadband interactivity*. That is, the telephone and computer networks have always been inter-active, but emerging information services and interactive television are creating a new function at the reception end of broadband telecommunications: user/receiver interaction and video return transmission. Even the smallest user, the individual in his/her home, can send movies upstream.

This interactivity creates a new dynamism in the nature of information and its relation to the communications process.

That is, in the earlier stages, power and intelligence resided first at the center and moved downstream. It then moved steadily toward the extremities to the point that intelligence in the nodes, the terminal equipment, or even off-network is now common. As we approach Stage 3, the power and intelligence are turning upstream. The user has control and ownership of the remote control, the telephone, the television set, and the computer terminal. Congress has turned its attention now to the new set-top boxes that will interface with the next generation of video on demand. And I would suggest that the consumer's ability to own or lease inside wiring will and should extend to outside wiring—that is, the broadband drop to the curb. (Still at issue, however, is the interface between the drop and the fibers at the pedestal.)

As competing fibers to the curb are economically feasible, these electronic driveways demonstrate this new bidirection of power in the network. Similarly, software agents, as well as filters, send intelligence in both directions, adding multidirection and complexities to the communications networks of the next era.

Economic Trends

These technological developments and other economic trends are creating two overriding phenomena: (1) the *commoditization of information* (pay-per-minute "transactional television" and information services on computer networks) and (2) the paradoxical *convergence and fragmentation* of economic entities.

Communications networks are narrowing the continuum between advertising and consumption. That is, if today an entity advertises on television, it expects results at the store, or possibly by a telephone call. Home shopping networks and computerized shopping on on-line networks are moving the locus of the transaction to the telecommunications network. As interactive broadband networks are created, with video on demand, the ability to advertise and sell in the same process will be heightened.

Producers and distributors of information try to find multiple revenue streams for the supply of information. For example, programmers seek advertising and subscriber revenues for the same program, as cable television networks and commercial computer networks such as Prodigy currently receive. Cable television is tending toward a la carte network ordering and greater use of the pay-per-view channels; Prodigy has moved to a basic rate with add-on charges for extra services. Strategic alliances, acquisitions, and business marriages raise fears of new levels of concentration and vertical integration; at the same time, successful enterprises are arising from lean entrepreneurial ventures and cottage businesses.

Users will also likely increase their demand for communications services in many ways. For example, Edward Shortliffe in his paper, "The Changing Nature of Telecommmunications and the Information Infrastructure for Health Care," in this volume, highlights the increased demand likely from health care institutions as a result of changing policy initiatives in that sector.

Sociological and Organizational Trends

The convergence and fragmentation trends, in fact, are more than economic. They occur in broader contexts and account, I believe, for another force that warrants close scrutiny.

It is clear that the general populace is less trusting of its traditional intermediary institutions such as political parties and candidates, journalists, educators, religious leaders, and many other mediating institutions in the current social environment. Along with that distrust comes a technology that allows the citizen to bypass many of those intermediaries. With the new transactional telecommunications, one can work, learn, shop, bank, pray, and even vote at home. The ability to bypass the traditional gatekeeper, to "disintermediate," coupled with increasing information overload, creates new pressures on older intermediary institutions to alter their gatekeeping functions by either adapting to the new information environment or becoming obsolete and giving way to new ones.

That is, there will remain a need for effective new or relegitimized old intermediaries. These *neointermediaries* will have to act as agents, filters, and integrators. They will connect people and information across space and time, serving the functions of knowledge navigation, information analysis, and system integration. These functions will move in all directions, working for the individual in an information-abundant world and as analyzers, information differentiators, and audience integrators for the entity trying to reach or target individuals. Changes in the functions of mediation may be the most significant of the nontechnological trends with respect to impact on the communications and information milieu.

Political Trends

Despite the movement against politics as usual, against taxes, and against big government, citizens are demanding greater meaning from their governments and political institutions. There is a longing on both the left and the right for a return to traditional democratic values and, with some, to orthodoxy. Some would characterize our current era's "green movement" and its attention to sustainable development as an indication of the need to recede from the model of the marketplace as the operative religion of the world's political and economic spheres.

In the communications regulatory realm, passage of the cable television reregulatory legislation over the veto of then President Bush, regardless of how one looks at the merits of the act, was an indication that purely deregulatory, competitive solutions were no longer politically sufficient for the country's problems. In the communications field, I would suggest, the regulatory regime will have to address other core values besides efficiency. These would include liberty, equity, community, and participatory access, if not others as well. The methods for doing so, on the other hand, will most likely have to be different from either of the preceding paradigms.

Scientific Trends

While one might consider scientific trends in the same breath as technological, they are not the same. The new sciences of chaos and complexity, drawing on teachings from quantum physics, modern evolutionary theory, ecology, and some of the newer thinking in economics, could be instructive to our thinking about new paradigms in regulation.[110]

One interesting approach to these topics is some recent work on coevolving complex adaptive systems. I will not expand on that here other than to suggest that in our exploration for a new paradigm for regulating the information infrastructure, we should consider what that model might teach us.[111] Specifically, it recognizes that there is a need to look at the crude whole rather than individual specialties in a vacuum. It suggests that systems coevolve, affecting each other as each adapts to the others. It recognizes that adaptive systems have memories, methods for prediction, feedback systems, and abilities to change to adapt to new conditions. It requires freedom for chaotic behavior but does not necessarily rely on invisible hands. Indeed, it bears mentioning that human organizations differ significantly from biological ones by the ability to apply intentionality: one can impact how something evolves in a number of ways.[112]

The science of complexity also addresses self-organizing systems. In the communications world the emergence of the Internet is an interesting example of a self-organizing, complex, adaptive system and one engendering a great deal of attention as a possible model for future networks.

The science of ecology is also forming the basis for broader thinking about human activity,[113] for example, the green movement, social ecology, and political references to sustainable development. In the next section I suggest some of the major issues we must address in the government's interaction with the production, distribution, and reception levels of the communications process. Perhaps they will show the way to a new communications landscape, conducive to a coevolution of technology, applications, and regulations to "sustain" our democratic value system.

Attention to scientific models is understandable in light of the current period of convergence and upheaval. The traditional boundaries of technology, of functionality, or regulatory regimes are dissolving. As Monroe Price suggests, "Convergence constitutes the abolition of given categories and the invention, perhaps, of new ones. That [is] what scientists . . . have done: question the existing categories and reconceptualize the world unburdened by the musty thinking of the scholastics."[114] These mergers, combinations, and convergences may transform the communications process and the nature of their regulation.

At the same time, we need to keep in mind our own ideology, a faith in the democratic process. Thus, as policymakers look at the evolving environment surrounding the information infrastructure, they will also want to act so as to foster democratic values.

TOWARD A NEW REGULATORY PARADIGM
FOR TELECOMMUNICATIONS REGULATION

Much of each of the regulatory schemes of Stages 1 and 2 is still with us today. Indeed, one proponent of the competitive paradigm suggests that the current regime is more regulatory than the past, in an effort to manage the transition to competition.[115] In any event, there is both extensive regulation currently and a strong consensus to move to a more competitive environment for the production, delivery, and reception of communications in the future. The question is this: Is the newer competitive paradigm the goal that has yet to be achieved, as Kellogg et al. (1992) suggest, or do we want to return to a more public-interest-oriented regulation, as others long for? Or should the country turn to a still newer approach to govern our most important political commerce, the delivery of information?

To answer that question, to find the regulatory religion of tomorrow, we have to have a vision not of the highway but of where the highway is headed. We have to have a better concept of the environment through which the highway is winding and where the highway leads. We need to complement our technological visions with scenarios of our society and of our democracy.

Furthermore, to have a vision of the society, we need to understand the nature of the coevolving systems surrounding the communications processes, which include, at the least, interactions among (1) the new technologies, (2) applications of those technologies, and (3) regulations impacting them. As each interacts, changes, and evolves, it affects each of the others in sometimes unpredictable ways.

For example, in the area of advanced or high-definition television, the technologies advanced during the 1980s from an analog model to a digital one. The applications for the technology thereby changed from better pictures to a much more flexible, scalable television, multiple channel delivery, and other uses converging television with computing. Furthermore, the regulatory process (e.g., spectrum allocation and standards setting) also evolved as the changes in technology and applications took shape. Finally, the technology progressed in a certain direction when the FCC

specified its criteria for the new advanced television system. Each of these areas, then, is impacted by the other two and by other influences as well.

It is important, therefore, to recognize this very dynamic process and to adopt intelligent policies, in that light, to foster societal and democratic values. Policymakers in the governmental, business, and nonprofit worlds can then design buttons and levers to adjust the landscape so that these sectors grow and adapt in the general form desired.

Governmental policy tools can come in the form of financial and tax incentives; research and development grants; subsidies on the demand and/or supply side; governmental loans and purchases;[116] standards setting; direct structural or behavioral regulation at the federal, state, or local levels; and perhaps new forms or methods for creating a proper atmosphere for democratic applications within the communications and information sectors.

These policies will certainly have to be flexible to allow the communications networks to self-organize, evolve, and adapt to new policies (such as health care initiatives) and other new demands. They will have to reflect the movement to a new regulatory stage, from scarcity to abundance, and now to complexity.

MAJOR ISSUES AHEAD

If the agenda for regulatory Stage 1 was essentially set by the lawyers (what is in the public interest?) and Stage 2 by the economists (what is efficient?), to whom do we turn for setting the agenda of Stage 3? My suggestion is that the scientists will help in providing models for analyzing the environment but not in setting the agenda for regulatory action. The new agenda setters for Stage 3 will likely be the political scientists (what is democratic?). That is, while regulators will want to allow growth and development in every sector of the telecommunications/information infrastructure, they will also need to judge and address them according to traditional democratic values. Like the environmental movement, in the face of rapid growth in the communications and information sectors, the new information environmentalists will want to assure "sustainable democracy," or the retention of democratic values and nondiminution of our most important human and information resources today and for future generations.

While this paper cannot suggest the specific policies to adopt at this time, it can suggest some of the underlying values and approaches we have traditionally wanted in an information infrastructure to serve broader democratic goals. Correlating to the values of efficiency, liberty, equality, community, and participatory access, the approaches might include:

- Creation of an efficient, reliable, adequate, flexible, and accessible infrastructure;
- Promotion of communications and privacy rights and responsibilities;
- Recognition of intellectual property interests;
- Concern for equality and equity, particularly with respect to an evolving concept of universal service; and
- Maintenance of public space for community discourse and as a public resource, with diverse and equitable access to that space.

Following are some key issues for consideration in designing a new approach to foster these goals and values in the networked telecomputing society of the next decade. Under each element of the communications process, I briefly address what I consider to be the three most significant issues in creating a communications landscape to sustain a democratic value system.

Production

The three most significant issues in the production realm are (1) the regime for intellectual property, (2) the use of government to enforce content controls, and (3) the creation of public space and governmental support for programming and software.

Intellectual Property

The first issue is perhaps the most vexing of the legal issues of the new age. It places the needs and interests of creators, authors, and inventors to protect and exploit their works, on the one hand, against the interest of society in general to have access to a wide diversity of information. This tension, it could be argued, places the values of liberty, property, and efficiency against those of equality, community, and participation. With digitization and convergence, the issues become blurred in several respects.

First, copyright is premised on reducing an expression to tangibility. The new uses of the information infrastructure enable works to be fluid. Potentially, they can be manipulated, reordered, and commingled from diverse sources. Intellectual property laws of the future will have to contend with the need for attribution of authorship, compensation of partial authors, and the inability to control distribution channels of a work.

Second, multimedia programs are information gluttons, calling for thousands of works in order to be true and effective. For example, should a historical multimedia program contain only the writings and pictures that are available through the public domain or through contractual agreement? If so, as the current copyright law would appear to require, the program itself may have to distort the nature of the historical point. For example, recent disputes over the ownership of Martin Luther King, Jr.'s, papers could result in a multimedia program on the civil rights movement to be largely devoid of Dr. King's original works. While such a result is expected in the current world of intellectual property rights, it could distort the teaching tools of the next generation of students.

These and other similar issues in intellectual property suggest that authors and major users will have to work out some form of group or agency licensing, technological attribution system, compulsory licensing, or other method of compensating authors, perhaps without their complete control of the distribution of their works.

Content Controls

With respect to content controls, the proliferation of programming would suggest that some new programming reaching the market will be beneath the current standards of good taste and decency (in terms of sexual content or violence). With the consumer's ability to control programming on the receiving end, however, the need to suppress speech at the production end should be minimized. Thus, with the abundance of sources at the production end and additional distributional outlets, there is likely to be greater amounts of objectionable programming. But with the attendant movement toward deregulation and less scarcity, there is less the government can or will do about it. On the other hand, there should be more options available to the consumer to control information at the terminal end.

Public Space and Information

The third issue is the need for and willingness of the government to support communities (geographical and communities of interest) by the creation of public spaces and substantive products (programming, software) for consumption within that public space. This issue is also thought of as the next manifestation of the public broadcasting, or public telecommunications, concept.

Traditionally, libraries allowed the public access to information that was otherwise available only to those who could afford to purchase each book. What are the libraries in the future age of commoditized information bits? How does one connect to them, as opposed to accessing the physical collections of public book libraries? Who pays and where does the information get delivered?

The future of the public broadcasting system is also an issue. As the distribution systems explode, will the market fulfill the programming needs heretofore met by public broadcasting? How do we measure market failures for public service programming? Should programming be subsidized? If so, which kinds? Who creates it? For what media? The current public broadcasting system is, of course, centered on the broadcast medium, where broadcasters are also producers. What should the future public telecommunications system look like, at what cost, and to whom? To what extent should the system extend beyond the broadcasting/video media?

Distribution

The issues for the distribution system revolve around the creation of an adequate communications infrastructure, the ability to gain access to the distribution system, and the impact that the infrastructure has on communities. The first issue addresses the value of efficiency, which today is usually equated with policies for competition. These policies, however, will also have to consider the values of equality and community, for the public is unlikely to tolerate a system that does not allow all to participate or that atomizes the country to the detriment of our community institutions.

Infrastructure

In the expanding global information environment, the United States already has a varied, diverse, and complex communications system that combines switched narrowband, broadband, wire, fiber, and radio paths to the end user. From this system to the vision of a nationwide (indeed, worldwide) switched broadband interactive video-on-demand system will take significant planning, patience, and determination. No doubt the U.S. communications and information highway is headed in that direction with only limited government involvement. But the public will have to take an active role in helping to shape significant elements of that distribution system. The regulatory principles of free flow of information, interconnection, access, diversity, and interoperability appear to be central to the democratic development and efficient evolution of the new international information environment. At this point in time, the evolution of the Internet—perhaps the most interesting and significant development of the 1990s in this field—has seen a push by government in the initial research and development, subsidies, and, arguably, in creating incentives but little in the way of direct regulation. As the government begins to pull away from its subsidization, and as access to the Internet becomes more prevalent and important, these issues should become more acute and visible.

Universal Service

Furthermore, social and governmental institutions will have to determine what "universal service" means beyond regular dial tone. How should that definition change over time and in different environments? What items should be considered essential and subject to controls to assure that most people have access? The public will have to determine how much it wants to invest directly in a telecommunications system that should (but may not) benefit the country economically and socially. What do we want to see the system do, such as provide educational, health care, citizenship, and cultural services? To whom and where should those services be delivered, and at what cost to whom?

Community

How do we design a system that will allow all citizens a form of editorial autonomy in the transmission and reception of communications and still have a society that recognizes value in community (however defined), in cohesiveness, in participatory democracy, and in equity? The network of networks comprising the information infrastructure will coevolve with the advance of telecommunications technologies and the software applications for which the distribution system will be used. The government's very difficult role will be to impact these applications in a positive direction while still allowing the flexibility for growth, adaptability, and sustainability.

Reception

Finally, we come to the issues associated with the reception end of the communications process. Here we might focus on the individual's interest in liberty and participation. That suggests the creation of an environment that empowers the individual to use the infrastructure, with the attendant benefit that the infrastructure will advance as more individuals become literate and intimate with it.

User Control

The new environment surely will promote the ability of users to exert control over the communications and information coming into the terminal location. Although the equipment industry is essentially private and subject only to competitive and safety restraints, regulation may be appropriate to empower the user to have a choice of equipment and of content over the network. This may take the form of violence chips, touch-tone phones, or signaling system 7 compatible equipment. More importantly, it may take the form of protecting against dominance of the set-top device that allows one to navigate the myriad information sources available on the new electronic networks, or even extending the residental users' ownership of wiring to the curb.

Privacy

Similarly, government will likely need to remain involved in the protection of individuals' privacy interests. But the extent of those privacy rights (as opposed to comodified privacy protections) is unclear. What are the bundle of privacy rights that must be guaranteed to every individual? Do they change in different environments? Who bears the costs in each situation?

Information Literacy

Finally, in order for the individual to be able to use any of the telecommunications goods and services, he or she will need to be information literate. This is the ability of a citizen to access, analyze, and produce information for specific outcomes (Aufderheide, 1993); access to the communications process will have to include access in terms of literacy as well as facilities and cost. This translates to the need for literacy education and awareness, along with the development of user-friendly technologies, which could be encouraged by government grants and purchases.

CONCLUSION

In conclusion, technological, economic, political, scientific, and organizational changes are ushering in a new age of complexity. The self-organizing interconnectedness of the Internet, the reverse direction of power in the network, and the convergence and dissolution of boundaries are bringing on a difference in kind more than of degree.

The future regulatory system for the nation's information infrastructure will need to go beyond the regulatory-deregulatory paradigms of the past and present. It will need to arrive at a new conception of the roles of governments in the evolution of the communications and information environment.

The new regime might look to ecological or scientific models to allow the flexibility of the various players to evolve and adapt to each other, to encourage or allow for the creation of neo-intermediaries, and to relate to the outside world. More specifically, technologies, applications, and regulations will all coevolve to form the new information and communications landscape of tomorrow. Governments will be charged with assuring that that landscape retains its democratic properties for current and future generations.

As U.S. society faces this process, it will have to balance the traditional values that are inherent in a democracy's political system: liberty, equality, community, efficiency, and participatory access. Unless policymakers assess and address that which is undemocratic in the new information environment, these values could be endangered. Under a new paradigm that fosters these strong values within the evolving international information environment, we can best attain a fair, efficient, and sustainable balance of private and public goals.

ACKNOWLEDGMENTS

The author acknowledges and thanks Marjory Blumenthal, Catherine Clark, Cameron Graham, Monroe Price, and Tracy Westen for their constructive comments and editorial suggestions on earlier drafts of this paper.

REFERENCES

AT&T Annual Report, 1909.

Aufderheide, Patricia. 1993. *Media Literacy: A Report of the National Leadership Conference on Media Literacy, the Aspen Institute Wye Center, Queenstown, Maryland, December 7-9, 1992.* Communications and Society Program, Aspen Institute, Washington, D.C.

Barnouw, Erik. 1975. *Tube of Plenty: The Evolution of American Television.* Oxford University Press, New York. (Second edition, 1990.)

Bollier, D. 1993. *The Information Evolution: How New Information Technologies Are Spurring Complex Patterns of Change.* Aspen Institute, Washington, D.C.

Bollier, D. 1994. *The Promise and Perils of Emerging Information Technologies: A Report on the Second Annual Roundtable on Information Technology.* Aspen Institute, Washington, D.C.

Brandeis, Louis D., and Charles Warren. 1890. "The Right to Privacy," *Harvard Law Review* 4:193.

Brenner, D. 1988. "Cable Television and the Freedom of Expression," *Duke Law Journal* 329.

Brenner, D. 1991. "What About Privacy in Universal Telephone Service?" *Universal Telephone Service: Ready for the 21st Century?* Annual Review of Institute for Information Studies (Northern Telecom Inc. and the Aspen Institute), Queenstown, Md.

Brenner, Daniel L. 1992. *Law and Regulation of Common Carriers in the Communications Industry.* Westview Press, Boulder, Colo.

Bruce, Robert R., Jeffrey P. Cunard, and Mark D. Director. 1986. *From Telecommunications to Electronic Services: A Global Spectrum of Definitions, Boundary Lines, and Structures.* Butterworths, Boston.

Chiron, Stuart Z., and Lise A. Rehberg. 1986. "Fostering Competition in International Communications," *Federal Communications Law Journal* 38(1):1-57.

Clarke, A. 1945. "Extraterrestrial Relays: Can Rocket Stations Give Worldwide Radio Coverage?" *Wireless World*, Vol. LI(Jan.-Dec.).

Cleveland, Harlan. 1993. *Birth of a New World: An Open Moment for International Leadership.* Jossey-Bass, San Francisco.

Dordick, Herbert S. 1986. *Understanding Modern Telecommunications.* McGraw Hill, New York.

Duggan, E. 1992. "The Future and Public Broadcasting," *The Aspen Quarterly* 4(3):14ff.

Entman, Robert M. 1992. *Competition at the Local Loop: Policies and Issues.* Aspen Institute, Washington, D.C.

Geller, Henry. 1991. *Fiber Optics: An Opportunity for a New Policy?* Annenberg Washington Program, Communications Policy Studies, Northwestern University, Evanston, Ill.

Information Infrastructure Task Force (IITF). 1993. *The National Information Infrastructure: Agenda for Action.* Information Infrastructure Task Force, Washington, D.C., September 15.

Kellogg, Michael K., John Thorne, and Peter W. Huber. 1992. *Federal Telecommunications Law.* Little, Brown, Boston.

Krasnow, Erwin G., Lawrence D. Longley, and Herbert A. Terry. 1982. *The Politics of Broadcast Regulation.* 3rd Ed. St. Martin's Press, New York.

Lawrence, John Shelton, and Bernard Timberg. 1989. *Fair Use and Free Inquiry: Copyright Law and the New Media.* 2nd Ed. Ablex Publishing Corporation, Norwood, N.J.

Levy, Steven. 1992. *Artificial Life: The Quest for a New Creation.* Pantheon Books, New York.

Lewin, Roger. 1992. *Complexity: Life at the Edge of Chaos.* MacMillan Publishing Company, New York.

Marx, Gary T. 1991. "Privacy & Technology," *Whole Earth Review* (Winter, 73):90-95.

McCarthy, J. Thomas. 1991. *The Rights of Publicity and Privacy.* C. Boardman, New York.

National Cable Television Association (NCTA). 1993. *National Cable Television Developments.* Washington, D.C., November.

National Telecommunications and Information Administration (NTIA). 1988. *NTIA Telecom 2000: Charting the Course for a New Century.* U.S. Government Printing Office, Washington, D.C.

Negroponte, N. 1993. "The Bit Police: Will the FCC Regulate Licenses to Radiate Bits?" *Wired* May-June, p. 112.

Nimmer, Melville B., and David Nimmer. 1978. *Nimmer on Copyright: A Treatise on the Law of Literacy, Musical and Artistic Property, and the Protection of Ideas.* 1993 supplement. Matthew Bender, New York.

O'Toole, James. 1993. *The Executive's Compass: Business and the Good Society.* Oxford University Press, New York.

Owen, Bruce, and Steven S. Wildman. 1992. *Video Economics.* Harvard University Press, Cambridge, Mass.

Pepper, Robert. 1993. "Broadcasting Policies in a Multichannel Marketplace," *Television for the 21st Century: The Next Wave,* Charles Firestone, ed. Aspen Institute, Washington, D.C.

Price, M. 1990. "Congress, Free Speech, and Cable Legislation: An Introduction," *Cardozo Journal of Entertainment and Communications Law* 8:225.

Prigogine, I., and Isabelle Stengers. 1984. *Order Out of Chaos: Man's New Dialogue with Nature.* Bantam Books, New York.

Reidenberg, J. 1992a. "The Privacy Obstacle Course: Hurdling Barriers to Transnational Financial Services," *Fordham Law Review* 60:§137.

Reidenberg, Joel R. 1992b. "Privacy in the Information Economy: A Fortress or Frontier for Individual Rights?" *Federal Communications Law Journal* 44(2):195-243.

Reuben-Cooke, W. 1993. "Rethinking Legal and Policy Paradigms for Television in the 21st Century," *Television for the 21st Century: The Next Wave,* Charles Firestone, ed. Aspen Institute, Washington, D.C.

Schmidt, Benno C., Jr. 1976. *Freedom of the Press vs. Public Access.* Praeger, New York.

Soma, John. 1983. *Computer Technology and the Law.* McGraw-Hill, New York.

Twentieth Century Fund Task Force on Public Television. 1993. *Quality Time? The Report of the Twentieth Century Fund Task Force on Public Television.* Twentieth Century Fund Press, New York.

Vogel, Harold L. 1990. *Entertainment Industry Economics: A Guide for Financial Analysis.* 2nd Ed. Cambridge University Press, New York.

Waldrop, M. Mitchell. 1992. *Complexity: The Emerging Science at the Edge of Order and Chaos.* Simon & Schuster, New York.

NOTES

1. See, for example, Bruce et al. (1986), which suggests several dichotomies such as basic versus enhanced, and facilities versus services. See also IITF (1993) (hereinafter, *Agenda for Action*), which suggests a broader definition that includes regulatory standards and people, as well as hardware, information, and applications.

2. See, for example, *National Broadcasting Company v. United States,* 319 U.S. 190 (1943) (Communications Act of 1934 is constitutional; FCC is more than just a traffic cop of the airwaves).

3. See *HBO v. Federal Communications Commission,* 567 F.2d 9 (D.C.Cir.), cert. denied, 434 U.S. 829 (1977).

4. Actually, the issue of privacy enters at each stage of the communications process. Privacy as the disclosure of intimate facts, for example, is properly viewed as a production-level issue and, to a certain extent, privacy could also be considered a distributional issue. For simplicity, however, this paper considers privacy issues as part of "reception."

5. Four of these values come from the work of James O'Toole (see particularly, O'Toole, 1993). The fifth value, participation, is the author's.

6. See, for example, 47 U.S.C. 326; Fairness Doctrine Report, 102 FCC2d 143 (1985).

7. See *National Citizens Committee for Broadcasting v. Federal Communications Commission,* 436 U.S. 775 (1978).

8. See, for example, 47 U.S.C. 312(a)(7), 315(a); Cable Access Rules, 47 U.S.C. §§531-32.

9. See Brenner (1991) and Marx (1991), pp. 90-95.

10. 47 U.S.C. 151.

11. See Statement of Senator Robert Kerrey, U.S. Senate, introducing S. 626, "The Electronic Library Act of 1993," March 22, 1993, in *Congressional Record.*

12. Primer on Ascertainment of Community Problems by Broadcast Applicants, 27 FCC2d 650 (1971).

13. *Henry v. Federal Communications Commission,* 302 F.2d 191 (D.C.Cir.), cert. denied, 371 U.S. 821 (1962).

14. See *Federal Communications Commission v. Midwest Video Corporation,* 440 U.S. 689 (1979).

15. *United States v. Midwest Video Corporation,* 406 U.S. 649 (1972).

16. See generally, Twentieth Century Fund Task Force on Public Television (1993) and Duggan (1992).

17. 47 U.S.C. 151.

18. Communications Act of 1934, Title II, 47 U.S.C. §§201-14.

19. *National Broadcasting Company v. United States,* 319 U.S. 190 (1943).

20. Communications Act of 1934.

21. See *United States v. Paramount Pictures,* 334 U.S. 131 (1948).

22. Telephone equipment, at the periphery of the communications process, could be used in the initiation or reception of communications messages. I consider it in the "Production" section of this part of the paper, because it is used first in the initiation of messages. The same is true of computer manufacturing, discussed below.

23. Complaint, *United States v. Western Electric Company,* No. 17-49 (D.N.J. Jan. 14, 1949). See Kellogg et al. (1992) at §4.3.

24. *United States v. Western Electric Company,* 1956 Trade Cas. (CCH) ¶68,246 (D.N.J. 1956), discussed in Kellogg et al. (1992) at §4.3.

25. In the later "Modified Final Judgment," agreeing to the divestiture of the Bell Operating Companies (BOCs) from AT&T, the manufacturing of equipment went with the competitive business and was forbidden fruit to the BOCs. See *United States v. AT&T,* 552 F.Supp. 131 (D.D.C. 1982), aff'd sub nom. *Maryland v. United States,* 460 U.S. 1001 (1983).

26. *National Broadcasting Company v. United States,* 319 U.S. 190 (1943).

27. 47 U.S.C. 315(a).

28. 18 U.S.C. 1464.

29. Mayflower Broadcasting Corporation, 8 FCC 333 (1940).

30. Report on Editorializing by Broadcast Licensees, 13 FCC 1246 (1949).

31. *Committee to Save WEFM v. Federal Communications Commission,* 506 F.2d 246 (D.C.Cir. 1974)(en banc). But see *Federal Communications Commission v. WNCN Listeners Guild,* 450 U.S. 582 (1981)(upholding FCC deregulation of format changes).

32. Primer on Ascertainment of Community Problems by Broadcast Applicants, 27 FCC2d 650 (1971); *Office of Communication of the United Church of Christ v. Federal Communications Commission,* 359 F.2d 994 (D.C.Cir. 1966).

33. See *Community Access v. Federal Communications Commission,* 737 F.2d 74 (D.C.Cir. 1984); *West Coast Media (KDIG) v. Federal Communications Commission,* 695 F.2d 617 (D.C.Cir. 1982), cert. denied, 464 U.S. 816 (1983) (failure to meet promises in public affairs and news).

34. 17 U.S.C. 102ff. See generally, *Sony Corporation of America v. Universal City Studios,* 464 U.S. 4117 (1984); Nimmer and Nimmer (1978; 1993 Supp.).

35. 17 U.S.C. 107.

36. 17 U.S.C. 111.

37. IBM Corporation, 687 F.2d 591 (2d Cir. 1982); Soma (1983) at §4.03; Kellogg et al. (1992) at §4.5, n. 7.

38. See Soma (1983) at §§4.02-4.18.

39. See AT&T (1909); Kellogg et al. (1992) at §1.3.

40. 47 U.S.C. 214.

41. 47 U.S.C. 201-214; Brenner (1992), pp. 15-18.

42. See *Louisiana Public Service Commission v. Federal Communications Commission,* 476 U.S. 355 (1986); Kellogg et al. (1992) at §§2.7.2 and 2.11.

43. See MTS and WATS Market Structure, Report and Order, 2 FCC Rcd. 2953 (1987); Brenner (1992), Chapter 9.

44. See Dordick (1986), Chapter 4.

45. INTELSAT Agreement.

46. See *Red Lion Broadcasting Corporation v. Federal Communications Commission,* 395 U.S. 367 (1969); *National Broadcasting Corporation v. United States,* 319 U.S. 190 (1943). But see *Federal Communications Commission v. League of Women Voters of California,* 468 U.S. 364, n. 11 (1984).

47. 47 U.S.C. 312(a)(7).

48. 47 U.S.C. 315(a).

49. 47 U.S.C. 315(b).

50. Report on Editorializing, supra.

51. Primer on Ascertainment of Community Problems by Broadcast Applicants, 27 FCC2d 650 (1971).

52. *Black Broadcasting Coalition of Richmond v. Federal Communications Commission.*

53. See *Community Access v. Federal Communications Commission,* 737 F.2d 74 (D.C.Cir. 1984); *West Coast Media (KDIG) v. Federal Communications Commission,* 695 F.2d 617 (D.C.Cir. 1982), cert. denied, 464 U.S. 816 (1983) (failure to meet promises in public affairs and news).

54. Prime Time Access Rule, 47 CFR §73.658(k).

55. 18 U.S.C. 1464.

56. *Federal Communications Commission v. Pacifica Foundation,* 438 U.S. 726 (1978).

57. 18 U.S.C. §1304 (transmission of lottery information prohibited), §1307 (state lotteries exempted). See also *United States v. Edge Broadcasting Company,* 113 S.Ct. 2696 (1993).

58. 18 U.S.C. §1343 (transmission of fraud prohibited).

59. See *Citizens Communications Center v. Federal Communications Commission,* 447 F.2d 1201 (D.C.Cir. 1971) at n. 9. See generally, Krasnow et al. (1982), Chapter 7.

60. *Red Lion Broadcasting Corporation v. Federal Communications Commission,* 395 U.S. 367 (1969).

61. *Miami Herald v. Tornillo,* 418 U.S. 241 (1974). For a discussion of the contrast between *Tornillo* and *Red Lion,* see Schmidt (1976).

62. *City of Los Angeles v. Preferred Communications,* 476 U.S. 488 (1986).

63. Cable Act of 1984, 47 U.S.C. §§521ff.

64. See 47 CFR §21.1ff; *National Association of Regulatory and Utility Commissioners v. Federal Communications Commission,* 525 F.2d 630 (D.C.Cir.), cert. denied, 425 U.S. 992 (1976).

65. See generally, Geller (1991).

66. Purchases by libraries may affect the production of certain books, tapes, and disks in that without this market certain academic works might not be published. Both the 1993 *National Performance Review* and the *Agenda for Action* (IITF, 1993) recognized the federal government's ability to impact the directions of the infrastructure through its purchasing power.

67. *Hush-A-Phone v. United States,* 238 F.2d 266 (D.C.Cir. 1957).

68. *United States v. Southwestern Cable Company,* 392 U.S. 157 (1968).

69. *Fortnightly Corp. v. United Artists Television,* 392 U.S. 390 (1968); *TelePrompTer v. CBS,* 415 U.S. 394 (1974).

70. See generally, *Malrite T.V. of New York v. Federal Communications Commission,* 652 F.2d 1140 (2d Cir. 1981), cert. denied, 454 U.S. 1143 (1982) (affirming later deregulation).

71. See Barnouw (1975), p. 100.

72. All Channel Receiver Act, 47 U.S.C. 330. See Krasnow et al. (1982), Chapter 6.

73. *Federal Communications Commission v. Pottsville Broadcasting Company,* 309 U.S. 134 (1940).

74. See also Vogel (1990), Chapter 1, which shows increasing share of nontheatrical revenues.

75. *United States v Columbia Pictures Industry,* 507 F.Supp. 412 (S.D.N.Y. 1980), aff'd F.2d (2d Cir. 1981).

76. Copyright Act of 1976, Public Law No. 94-533, 90 Stat. 2541, Oct. 19, 1976, at new Section 107, 17 U.S.C. §107. See Lawrence and Timberg (1989).

77. Cable Television Consumer Protection and Competition Act of 1992 (Public Law 102-385), Section 6, 47 U.S.C. §325 (b).

78. Specialized Microwave Common Carriers, 29 FCC 2d 870 (1971), aff'd sub nom. *Washington Utilities and Transportation Commission v. Federal Communications Commission,* 512 F.2d 1157 (9th Cir.), cert. denied, 423 U.S. 826 (1975).

79. Above 890 MHz, 27 FCC 359 (1959), recon. denied, 29 FCC 825 (1960).

80. Microwave Communications, Inc., 18 FCC 2d 953 (1969).

81. Specialized Microwave Common Carriers, 29 FCC 2d 870 (1971), aff'd sub nom. *Washington Utilities and Transportation Commission v. Federal Communications Commission,* 512 F.2d 1157 (9th Cir.), cert. denied, 423 U.S. 826 (1975).

82. *MCI Telecommunications Corporation v. Federal Communications Commission,* 561 F.2d 365 (D.C. Cir. 1977), cert. denied, 434 U.S. 1040 (1978); *MCI v. FCC,* 580 F.2d 590 (D.C.Cir.), cert. denied, 434 U.S. 790 (1978).

83. Establishment of Domestic Communications-Satellite Facilities by Non-Governmental Entities, 22 FCC 2d 86 (1970) ("Open Skies"); Establishment of Domestic Communications-Satellite Facilities by Non-Governmental Entities, 35 FCC 2d 844, 37 FCC 2d 184, recon. denied, 38 FCC 2d 665 (1972), aff'd sub nom. *Network Project v. Federal Communications Commission,* 511 F.2d 786 (D.C.Cir. 1975). See generally, Note, "The Satellite Competition Debate: An Analysis of FCC Policy and an Argument in Support of Open Competition," *Syracuse Law Review* 40:867 (1989), cited in Kellogg et al. (1992) at §12.3.3, n. 60.

84. Resale and Shared Use, 62 FCC 2d 588 (1977), aff'd sub nom. *AT&T v. Federal Communications Commission,* 572 F.2d 17 (2d Cir.), cert. denied, 439 U.S. 875 (1978).

85. Establishment of Satellite Systems Providing International Communications, 101 FCC 2d 1046 (1985). See Chiron and Rehberg (1986) and Kellogg et al. (1992) at §§15.3.3 -15.3.4.

86. Use of Bands 825-845 MHz and 870-890 MHz for Cellular Communications Systems, 86 FCC 2d 469 (1981).

87. Personal Communications Service (FCC General Docket 90-314, Sept. 23, 1993).

88. Expanded Interconnection with Local Telephone Company Facilities, CC Docket No. 91-141, Report and Order and Notice of Proposed Rulemaking, released October 19, 1992.

89. See NCTA (1993), p. 11-A. See generally, Pepper (1993).

90. National Cable Television Association (1993), p. 7-A (78 national cable video networks as of December 31, 1992). Added to those are the four national television broadcast networks: ABC, CBS, Fox, and NBC. Several new cable video networks were begun in 1993, including The Food Network and FX-TV.

91. *City of Los Angeles v. Preferred Communications,* supra.

92. *The Chesapeake and Potomac Telephone Company of Virginia v. United States,* Civil Case No. 92-1751-A (D.E.Va.) (Slip Op. of Judge T.S. Ellis III, August 24, 1993).

93. See, for example, Price (1990). Professor Price questions whether the First Amendment is being extended too far in these kinds of cases. See generally, Brenner (1988).

94. *Carterfone v. AT&T,* 13 FCC 2d 420, recon. denied, 14 FCC 2d 571 (1969).

95. Computer Inquiry II, 77 FCC 2d 384 (1980), recon. FCC 2d 445 (1981), aff'd in part sub nom. *Computer & Communications Industry Association v. Federal Communications Commission,* 693 F.2d 198 (D.C. Cir. 1982), cert. denied, 461 U.S. 938 (1983).

96. Id. See 47 C.F.R., Part 68. See New or Revised Classes of Interstate and Foreign Message Toll Telephone Service (MTS) and Wide Area Telephone Service (WATS), First Report and Order, 56 FCC 2d 593 (1975).

97. In the later "Modified Final Judgment," agreeing to the divestiture of the Bell Operating Companies (BOCs) from AT&T, the manufacturing of equipment went with the competitive business and was forbidden fruit to the BOCs. See *United States v. AT&T,* 552 F.Supp. 131 (D.D.C. 1982), aff'd sub nom. *Maryland v. United States,* 460 U.S. 1001 (1983).

98. For example, New York Civil Rights Law, §§50-51. See also, McCarthy (1991), pp. 6-4 and 6-5.

99. See Brenner (1991). See also NTIA (1988), pp. 135-140.

100. Fair Credit Reporting Act, 15 U.S.C. 1681.

101. Rights to Financial Privacy Act, 12 U.S.C. 3401.

102. Cable Communications Policy Act of 1984, 47 U.S.C. 551.

103. For example, Family Educational Rights and Privacy Act, 15 U.S.C. 1681; Video Privacy Protection Act, 18 U.S.C. 2710.

104. In that sense, the issues of privacy are not strictly within the "reception" category, but for ease of reading, the discussion of privacy is primarily in this section and not spread throughout the paper.

105. See, for example, Brenner (1991).

106. *Federal Communications Commission v. Pacifica Foundation,* 438 U.S. 726 (1978).

107. Indecency differed from obscenity in that it had to meet only one test—to be patently offensive to community values (for the nation as a whole, according to the FCC)—whereas obscenity had a three-part test that included an appeal to prurient interests and the work taken as a whole was without serious literary, artistic, political, or scientific value.

108. See, for example, *Cruz v. Ferre,* 755 F.2d 1415 (11th Cir. 1985). The courts also restricted federal approaches to regulating "indecent" content of sexually oriented 900 numbers. See, for example, *Sable Communications of California v. Federal Communications Commission,* 492 U.S. 115 (1989), but the courts have allowed regulations in this area that provided for access codes, scrambling of messages, and use of credit cards. See *Carlin Communications v. Federal Communications Commission,* 837 F.2d 546 (D.C.Cir.), cert. denied, 488 U.S. 924 (1988). Nevertheless, the courts have not sided entirely with vendors on this matter, and some restrictive regulations, even on carriers, were allowed [Mountain States].

109. Some could conclude that these vestiges are simply problems of transition, or regulatory noise. In light of the later discussion, I would suggest that often they are expressions of values so strongly held that even the newer paradigm would not release the regulatory need for more than a competitive response.

110. For descriptions of the movement toward a science of complexity, or on self-organization, see, for example, Prigogine and Stengers (1984), Lewin (1992), Levy (1992), and Waldrop (1992).

111. See Bollier (1993), pp. 4-6.

112. See, for example, Bollier (1994).

113. See, for example, Cleveland (1993).

114. Letter of Monroe Price to author, September 9, 1993.

115. Kellogg et al. (1992) at §3.

116. The workshop for which this paper was written explored various investment and incentive options. See particularly, Walter S. Baer, "Government Investment in Telecommunications Infrastructure," in this volume. For another thorough listing of government incentives, see IITF (1993).

Current and Future Uses of Information Networks

Colin Crook

This paper describes the uses of information networks, today and in the future. Citicorp is used as a specific example, but many of the comments are applicable across the financial and banking communities.

BACKGROUND

Citicorp offers banking and financial services in more than 90 countries. Products and services serve both individual consumers (approximately 40 million) and corporations, on a global basis. Information technology is fundamental to the operation of the bank. Indeed, money can be viewed as an ideal information technology product. In most instances, money exists as "bits" on computer storage disks, computer processors, networks, and so forth. The sheer volume of money movement per week across the bank's systems approximates the U.S. gross domestic product.

Central to the bank's structure is its communications network. The developments in information technology and distributed computing emphasize even more the increasing role of the network. The bank serves its consumer and corporate customers via electronic network means (customer automatic teller machines, or ATMs, and corporate mainframe connections). The network provides immense flexibility to the bank in serving its customers; locating its assets, independent of geography; delivering products globally; connecting with competitors and partners; and using other companies' resources to assist the bank.

The bank's networks in the United States are very extensive, providing voice, video, and data communications. The global reach of the bank requires that the extensive U.S. network is an integral part of the bank's global network. The communications network is rapidly becoming the bank's common infrastructure. This enormously complex network, which actually is a combination of many national network systems, must be viewed in the context of global networks. Today, however, no single company can yet offer a truly global networking capability. Citicorp can be viewed as an acid test of true end-to-end service delivery across the world.

THE IMPORTANCE OF INFORMATION NETWORKS

The bank's goal of total relationship banking requires that customers anywhere in the United States can access a customer service person via telephone. In turn, this service person must be able to access all of the customers' information, regardless of location. For example, a service center in San Antonio offers Spanish and English language support for the whole United States, with plans in place to support overseas operations in the future. The service person can access customer information in New York, Nevada, and South Dakota, where computer centers exist. The network is fundamental to this capability. A network-based infrastructure would also allow the shifting of business functions to locations that can provide operating efficiencies. The bank is transferring its European credit card processing activities to Nevada. That will require routinely moving lots of information across the network between Europe and the United States, and doing so in a secure way.

In terms of the bank's basic operation, its networks are already essential to its very survival—the bank cannot function without the network. However, the full impact of contemporary technology on both the bank's own systems and the overall economy has not yet been felt! Indeed, early evidence and experience indicate that this may be profound, affecting both the very structure of the bank and the fundamental behavior of its customers.

A GOOD ECONOMIC BET

The banking industry contains large amounts of inefficiencies that may be intrinsic to other industries as well. Citicorp's push for a network-based infrastructure assumes that telecommunications is a good economic bet. We believe, in fact, that networking information technology is a very good economic bet.

However, companies cannot exploit technology networking unless it is fundamentally integrated into the way the companies actually do business. It involves a systems approach that requires a rethinking of the entire business.

In addition, considerable technological innovation is taking place. The creative use of these developments could represent significant competitive threats. However, these developments could also provide the bank with the capability to fundamentally restructure its business processes as well as permit dramatic improvements in serving customers. In examining the likely developments in technology, several themes appear to be evolving that will affect the bank.

LIKELY DEVELOPMENTS

Customer Interfaces

The emergence of an increasing panoply of innovative consumer and business information devices (screen phones, personal digital assistants, smart cards, etc.) represents opportunities for new products and services. In addition, some form of payment system will be needed in order to support the wide range of services being sold. These new devices will be connected via networks to the service providers.

In essence, network usage, which used to be dominated by the plain old telephone, will dramatically rise through the increasing use of innovative information devices. Already, considerable amounts of financial activities are carried out by electronic means, such as ATMs. Therefore, financial and banking companies will seek to exploit the network to provide more effective consumer and business access to their services via the multitude of new information devices. Customers will probably wish to use electronic means to transact their business with the bank.

Bank Structure

In addition to providing customer access, the network facilitates and enables significant new ways to structure enterprises. Issues of both time and geography can be handled more effectively. For example, the processing of information can be handled at specific points around the world and, via use of the network, information can be moved quickly and efficiently. Citicorp already processes information in the United States for non-U.S. operations of the company. This trend will continue.

As companies operate globally, their needs must be handled 24 hours a day. For example, Citicorp is providing 24-hour-a-day service coverage from the United States in support of its global customers' cash management needs. The new awareness of systemic risk and potential threats to the bank's operations permits the bank to exploit the network to provide geographical and time diversity and risk allocation. Provided that the network remains available, the bank's operations can continue despite calamitous events (fire, bombs, natural disasters, etc.). The ability of the network to facilitate the operating integrity of the company, enable efficiency, and eliminate unneeded redundancy will grow in importance as the company grows. The network must have the needed characteristics of capacity, resiliency, and reliability.

The Network as a Market

Electronic commerce is increasing rapidly as companies restructure their business processes and exploit new thinking in terms of how companies can more effectively work together. The U.S. gross domestic product of approximately $5.6 trillion per year [as of 1993] uses around 100 billion financial transactions to support it (cash, credit cards, etc.). Interestingly, the telecommunications network, in order to operate, uses approximately 100 billion financial transactions each year. Already the information economy has more financial transactions than the U.S. gross domestic product!

As electronic commerce increases, the financial transaction volume will rise exponentially. Electronic commerce carried out across networks will represent new challenges requiring considerable innovation. Already trillions of dollars are transferred by the bank every week—and most of this activity is performed not by people but by machines. Such volumes mandate a high level of bank control over access to its network-based applications. In a network-based business environment, protecting the application by authenticating application users is critical because of the extreme difficulty in protecting the network itself. The network will have literally millions and millions of customers—individuals, companies, public institutions. Because customers tend to do whatever they want, the key is to build the right kind of infrastructure that can support the needs of the individual customer.

Exponential rises in electronic commerce will also lead to pressures for new legal and regulatory thinking. Without such new thinking, business and commerce could be inhibited. Intellectual property issues are already proving to be thorny ones.

The Network Structure

A contemporary view of a network is to regard it as an information infrastructure with an array of essential network services that facilitate the flow of customer products and services across it. If the network is just viewed as bandwidth, it becomes very difficult to use. The services inside the network, such as security, directories, and so forth, are essential to its use by intelligent devices or software applications.

As enterprises become more network-based, they are seeking to have other companies supply capabilities that permit the enterprise to focus on those processes that create and deliver value to their customers. The network facilitates the creation of the so-called virtual enterprise. Here, directories are critical to permitting several companies to appear to be a single enterprise. Directories are central to the network.

Citicorp's future strategies anticipate the emergence of the virtual enterprise, and work is already in hand to move in this direction. The future network structure of the bank will be a combination of public and private networks. Wherever possible, the public network will be exploited. Additionally, the bank will be a combination of the capabilities of several companies, all existing within an integrated information network infrastructure. This vision of the bank will not be static. It will be constantly changing, according to customer and business needs.

Critical to the successful evolution of these ideas is the emergence of common infrastructural services that all companies will utilize. An example of this is a common, secure, trusted billing and payment infrastructure that many institutions can use. This would address issues of auditability, fraud prevention, user authentication, secure access control, and so forth. It is believed that progressive economies will develop such common information network-based capabilities that users and suppliers can use, obviating the need to develop multiple proprietary networks with no customer value.

The Changing Nature of Telecommunications and the Information Infrastructure for Health Care

Edward H. Shortliffe

INTRODUCTION

Discussions at this CSTB workshop and in other forums are confirming the notion, often voiced in medical computing circles, that the use of computers and communications for health care is roughly a decade behind routine applications of the technologies for much of the rest of society. The details about Citicorp provided by Colin Crook in this volume, however, suggest that the gap is closer to two decades! Much is happening in the medical computing world, but the story is different from that for many other segments of society, and here I will try to explain some of the reasons for those differences.

It would be inaccurate to give the impression that the medical community has been oblivious to the potential role of computers and networking in biomedicine. In fact, in the early 1970s the first node on the ARPANET that was not a defense-funded resource (either through U.S. Department of Defense grants to academic institutions or direct military support) was the National Institutes of Health-funded SUMEX-AIM resource at Stanford University's School of Medicine, a machine dedicated to biomedical applications of artificial intelligence research. With the subsequent addition of many more medically oriented nodes to the national network, a small but active segment of the biomedical community grew up with the ARPANET during its transition to the Internet of today.

Furthermore, it is revealing to look back to a prophetic 10-year-old document that was produced by a long-range planning panel for the National Library of Medicine (NLM) in 1986. Shortly after the new director of the NLM took office in 1985, he brought together people who could help devise a grand plan for what the institution should be doing to prepare for the decades ahead. The resulting report (NLM, 1986) dealt with the future of medical "informatics" and noted that "widely disseminated medical information systems will require high-bandwidth communications to allow access to the computational, data, and information resources needed for health care and research" (p. 60). An explicit goal mentioned in the report was that "by the end of the next decade [presumably 1996], there will be a national computer network for use by the entire biomedical community, both clinical and research professionals. The network will have advanced electronic

NOTE: Support of this work under grants LM-05305 and LM-05208 from the National Library of Medicine is gratefully acknowledged. Portions of this paper are adapted from a presentation given by the author at the 1993 annual meeting of the American Clinical and Climatological Association held in Sea Island, Georgia (Shortliffe, 1994). Many of the topics in this paper were also discussed at "National Information Infrastructure for Health Care," a workshop sponsored by the Computer Science and Telecommunications Board and the Institute of Medicine on October 5-6, 1993, in Washington, D.C.

mail features as well as capabilities for large file transfer, remote computer log-in, and transmitted graphics protocols. It will either be part of the larger national network . . . or will have gateways to other federally sponsored networks" (p. 65). The report includes much discussion of the ARPANET, as the larger Internet was still called in 1986, as a model for the national network that would facilitate a variety of applications in biomedicine and health care delivery.

Despite this explicit call for the biomedical community to embrace the potential role of a national communications infrastructure, little happened in the intervening years. The NLM expressed an interest in pursuing the topic, but it needed incremental funding to do much and the rationale for new efforts was extremely hard to sell to the mainstream medical community and hence to the Congress. Yet since the passage of the High Performance Computing and Communications Act and the election of an administration with particular interests in both the national information infrastructure and health reform, we have seen the awakening of interest among leaders for whom computers and telecommunications for health care had previously been viewed as an esoteric topic.

STIMULI TO CHANGE

Certain key forces are driving such changes in awareness and interest. First, the shift to managed care and capitation is changing dramatically the requirement for communication among the parties involved in health care. Insurers are demanding a basis for making comparisons among providers (both institutional and individual), and suddenly new kinds of clinical data need to be collected, communicated, and collated. Pressures to develop and manage such comparative clinical data did not exist in the past to the same extent.

Second, proposed health care reform legislation, and the resulting high-profile discussions of health care financing and organization, are reinforcing and broadening the pressures on health care institutions and providers. When President Clinton introduced his health reform proposals to the Congress on September 22, 1993, he referred explicitly to the opportunities for increased efficiency, technology assessment, and cost savings offered by information technology. Computing and communications technologies have emerged as key elements in the strategic plan for eliminating waste in the health care system. Clearly, such an impact will be easier claimed than achieved, but the expectations do help explain the sudden shift in interest among health planners.

There are other prominent examples of a growing societal awareness of the potential role for the national information infrastructure (NII) in supporting health reform. In April 1993, the Computer Systems Policy Project (CSPP), composed of senior executives from the major vendors of computer systems, released a report on the relevance of the NII to health care (CSPP, 1993). During that same month, a report commissioned by the U.S. Department of Health and Human Services, and developed by a panel formed by the American Hospital Association, was released (AHA/DHHS, 1993). Although the committee that drafted the report had been convened to look at issues in the creation of computer-based patient records, it soon chose to address more broadly the issues of information infrastructure required to support the notion of computerized individual patient records. This broader view is especially valid when one begins to envision longitudinal medical records that are tied to a mobile patient rather than to a single provider's office or a hospital.

For those who have been interested in medical informatics and biomedical communications for some time, now is an exciting period. Key decisionmakers are listening and becoming very enthusiastic about seeing profound changes, both in the health care system itself and in the creation and use of an underlying information infrastructure. Until recently, the role of regional and national communications in support of health care has been a largely grass-roots activity, with limited shared national vision and leadership. Some of the most successful experiments have been in the area of "telemedicine," in which, for example, electronic communications have been used to provide consultation by specialists to physicians in rural, inner-city, or other isolated locations. As the Internet increases in its capacity, it will be able to accommodate the kinds of voice and video transmissions that are crucial for this kind of telemedicine activity. Constraining progress, however, have been frequently voiced concerns regarding risks to the privacy and security of clinical information. Such concerns have led many health care institutions to resist exploring modern networking technologies, both within their own walls and when considering linkage to outside networks in their communities or beyond (IOM, 1994).

RECOGNIZING THE NEED FOR IMPROVED CLINICAL DATA SYSTEMS

Until 1993, there was essentially no federal involvement in defining the role of the information infrastructure as it relates to the delivery of health care. Beginning in the late 1980s, largely through the activities of the NLM, we did see the involvement of the health sector in discussions of how the NII might support research and education in biomedicine. Unfortunately, there has been little or no knowledge of the existing NII, nor an understanding of its implications, among the leaders in the health care industry. Those few hospitals that are connected to the Internet are mostly academic institutions that have sought such connections through their main university campuses. Recently, the NLM initiated a grants program to encourage more hospitals to institute Internet connections and to begin to explore the ways in which national networking could support their clinical mission.

Obstacles to more effective use of the existing NII in health care, and to an informed anticipation of how emerging communications and computing technologies will affect health care, are largely logistical, political, and financial, rather than technological. About two years ago I was asked to give a talk at a conference on gigabit networking. My message was that, for the present, we can largely ignore biomedical gigabit networking issues and simply work to make better use of the technologies that we have today. That is not to say biomedicine could not do more with gigabit speeds in the future, but that is not the major need at present.

One way to make progress in dealing with the logistical, political, and financial barriers to acceptance of computing and communications technology has been to demonstrate the relevance of such technology to cost savings and to health reform. We are beginning to see data in the literature that demonstrate how computing holds the promise of impressive economies. One recent report, from physicians at the Regenstrief Institute at the Indiana University School of Medicine, describes a well-designed controlled clinical trial demonstrating more than $800 in savings per patient stay at Wishard Memorial Hospital in Indianapolis during which some physicians used computers to order tests and to receive reminders, whereas other providers did not use the technology (Tierney et al., 1993). As the article notes, extrapolation of these effects suggests "savings of more than $3 million in charges annually for this hospital's medicine service and potentially tens of billions of dollars nationwide" (p. 379). The problem is that large complex systems such as the one built at the Wishard Memorial Hospital and connected to the Regenstrief Medical Record System over the past 20 years could not be duplicated simply for implementation at another hospital. In the absence of standards for systems integration and data sharing among institutions, transporting a highly tuned technology from one hospital to another can be next to impossible.

Data such as those from the Indiana study are clearly needed to demonstrate the value of interinstitutional network connections and the role that the NII could play. Such fiscal data make hospital CEOs pay attention, and they are having much to do with a reassessment of how hospital data systems need to be designed and especially how they might become more clinical in their emphasis, departing from a traditional administrative and financial orientation.

Also driving the need for more clinically oriented data systems is our current lack of data to gauge the quality or cost of health care. Employers and insurers are increasingly choosing to contract with the hospital that can show the highest quality at the lowest cost; if an institution charges more, it must be able to show that the higher cost is associated with higher quality. If provider institutions lack the data systems that allow them to demonstrate improved outcomes over competing hospitals, they may increasingly find that they are unable to win the managed care contracts required to keep their beds full. Subjective impressions that one hospital is "better" than another hold little sway with an employer or insurer that is fighting frantically to control health care costs.

There are generally no community-wide databases that store information on providers and patients, although there are a few experiments to develop regional health databases (IOM, 1994). When there are pooled data for regions, they tend not to record patient-centered information on topics such as consumer satisfaction or functional outcomes after treatment. Similarly, there is generally no reasonable way to determine whether doctors are performing safe, appropriate, and effective care, despite demands that we begin to develop the kind of data sets that allow such information to be released.

We clearly need better data collection methods than we have today. For example, clinical data sets derived from insurance claims and depending on voluntary submission of diagnostic and outcomes information by providers are often rendered useless by inconsistent compliance and information of questionable accuracy. Several recent studies regarding the health of our nation are based upon data submitted to Medicare on claims forms. Clinical information on such claims provides a limited proxy for medical reality. This helps explain the Medicare system's recent pressure for the creation of a Uniform Clinical Data Set (Audet and Scott, 1993). We clearly need better ways of collecting comprehensive clinical data than to depend on insurance claims submissions. This is one of the many justifications for the recent push to see the creation of computer-based patient records (IOM, 1991).

ATTRACTING PHYSICIANS TO INFORMATION MANAGEMENT TOOLS

A colleague and I recently argued that clinicians are inherently "horizontal" users of information technology (Greenes and Shortliffe, 1990). By this we mean that they require a critical mass of functionality from the system they are using before routine use of the computer will be viewed as worthwhile to them. If the computer is useful only occasionally, say for one or two patients per day, then the inertia involved in going to the machine will typically prevent the effective use of that technology. But if the computer provides functionality that is useful for essentially every patient seen, and if that use is as good as or better than the manual methods previously available, then it is reasonable to assume that physicians will begin to turn to the computer for support. It is also important to recognize that physicians seek help with the noxious tasks associated with data management but are not interested in having computers infringe on valued tasks. Furthermore, they require intuitive interfaces that require little or no training (similar, for example, to the training required to use the telephone or, more topically, an automated teller machine).

Part of this critical mass of functionality will be made available not within the physician's local environment but via the NII. Imagine the model shown in Figure 3, in which providers are linked, either directly or through the hospitals at which they practice, to research and public health databases or to national repositories of patient data. When people become sick away from home, there would be tremendous utility to a system that permitted authorized health workers, say in an emergency room, both to identify you and to access key clinical data about your medical problems, allergies, medications, and the like from a centralized data resource provided via the network. Physicians could also be provided with access to third-party payers, to Medicaid/Medicare, to the Food and Drug Administration, to the Centers for Disease Control, to medical schools and their

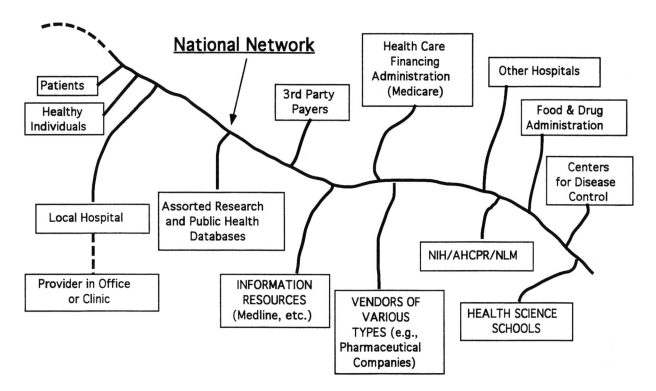

FIGURE 3 A view of the resources that could be made available to practitioners, patients, and the public via a national information infrastructure for health care.

continuing education activities, to a variety of vendors such as pharmaceutical companies or software companies, to information resources such as Medline, and to patients themselves. If the network provides health workers, patients, and the public with access to all these kinds of information sources, as well as two-way communication that allows retrieval of data plus submission of information (e.g., claims), one can begin to imagine the appeal and acceptability of the NII and its health care role. There has been remarkably little planning initiated regarding the implementation of these kinds of connectivity, but I believe that Figure 3 demonstrates that the NII has the potential for a major impact on health care.

EXAMPLE USES OF THE NII FOR HEALTH AND HEALTH CARE

A workshop sponsored jointly by the Computer Science and Telecommunications Board of the National Research Council and by the Institute of Medicine ("National Information Infrastructure for Health Care," October 5-6, 1993) offered a variety of possible uses of a national information infrastructure for health and health care. I close by summarizing them here since they follow naturally from the discussion above:

- Information distribution and access, including bibliographic-retrieval software for searching the medical literature;
- Population databases (regional, state, and national) with specialized interfaces that allow retrieval of subsets of patients meeting particular search criteria;

• Access to longitudinal, sharable, standardized health records for individual patients, particularly important for providing continuity of care for our highly mobile society;

• Telemedicine, especially to provide enhanced care and information access in underserved areas such as in rural regions or inner cities;

• Personal health information systems, which would provide individuals, whether sick or healthy, with educational materials plus a personally maintained health database;

• Databases for research and outcomes assessment, as previously described;

• Systems to handle billing, finance, reimbursement, and eligibility determination;

• Multimedia communication and video conferencing;

• Implementation of practice guidelines and outcomes management advice with specialized software at the point of care that allows access to the individual guidelines that may be available locally or over the network; and

• Submission of clinical reports to federal agencies, such as reportable disease information to the Centers for Disease Control or adverse drug reaction reporting to the Food and Drug Administration.

CONCLUSION

The future world that I have described here (assuming appropriate safeguards to protect patient privacy and confidentiality of data) offers a set of features that many observers believe would not only be acceptable to practitioners but also would enhance their practices in positive ways while helping to reduce some of the waste in our current health care system. Unlike other technologies that have played a role in escalating the cost of health care in this country, there is reason to believe that computing technology, coupled with a standardized communications infrastructure, could actually eliminate waste and reduce the total health bill. To achieve these goals, however, various enabling activities are required. Among these are the need for improved national leadership and a greater understanding of the federal role in guiding the development of standards, the education of practitioners and others regarding the role of the NII for health care, the creation of incentives, and attention to how the health care system should reimburse those who use the information infrastructure in support of health care delivery. Several observers have noted the need for preemptive federal legislation in the areas of privacy and confidentiality in particular, but also to deal with authentication of electronic signatures to assure their acceptability for legal documentation. The goal of developing centralized longitudinal lifelong medical records that could be accessed by providers (when authorized to do so by individual patients; see IOM, 1994) requires that we address the need for national patient and provider identifiers while balancing that need against the civil liberties issues involved in linking health care identifiers with other identifiers used in our society.

There is clearly a major need for training and education, not only of health professionals but also of the public. In addition, we need a new cadre of medical computing professionals who not only understand the technological issues involved but also have practical experience in clinical settings (if not formal training in one of the health professions) so that they are knowledgeable about the realities of health care practice, ethics, and financing and can incorporate resulting sensitivities into the systems that they design and build. Finally, there is a major need for demonstration projects to help prove the technology's cost-effectiveness and its impact on the quality of care. One complicating factor in the evolution of this field has been the difficulty in developing demonstration projects with sufficient scope, penetration, and generalizability to assure that they can provide meaningful data regarding the technology's potential impact on cost and quality.

REFERENCES

American Hospital Association (AHA)/U.S. Department of Health and Human Services (DHHS). 1993. *Toward a National Health Information Infrastructure.* Work Group on Computerization of Patient Records. American Hospital Association, Chicago, April.

Audet, A., and H. Scott. 1993. "The Uniform Clinical Data Set: An Evaluation of the Proposed National Database for Medicare's Quality Review Program," *Annals of Internal Medicine* 119:12091213.

Computer Systems Policy Project (CSPP). 1993. *Information Technology's Contribution to Healthcare Reform.* Computer Systems Policy Project, Washington, D.C., April.

Greenes, R.A., and E.H. Shortliffe. 1990. "Medical Informatics: An Emerging Academic Discipline and Institutional Priority," *Journal of the American Medical Association* 263:1114-1120.

Institute of Medicine (IOM). 1991. *The Computer-Based Patient Record: An Essential Technology for Health Care,* R. Dick and E. Steen, eds. National Academy Press, Washington, D.C.

Institute of Medicine (IOM). 1994. *Health Data in the Information Age: Use, Disclosure, and Privacy,* M.S. Donaldson and K.N. Lohr, eds. National Academy Press, Washington, D.C.

National Library of Medicine (NLM). 1986. *Long Range Plan on Medical Informatics.* National Library of Medicine, Washington, D.C., December.

Shortliffe, E.H. 1994. "Clinical Information Systems in the Era of Managed Care," *Transactions of the American Clinical and Climatological Association,* Vol. 105, pp. 203-215.

Tierney, William M., Michael E. Miller, J. Marc Overhage, and Clement J. McDonald. 1993. "Physician Inpatient Order Writing on Microcomputer Workstations: Effects on Resource Utilization," *Journal of the American Medical Association* 269(3):379-383.

Can K-12 Education Drive
on the Information Superhighway?

Robert Pearlman

If they build it, will we come? If government, the telephone and telecommunications companies, and the cable industry join to develop the backbone of the information highway and its local access ramps, will schools and school districts invest in the local telecommunications infrastructure that will ensure universal participation by the nation's over 40 million K-12 students and their teachers?

While the government's goal to "extend the 'universal service' concept to ensure that information resources are available to all at affordable prices" (IITF, 1993) may be a reasonable short-term *policy* for federal government action, it is at best only a first step toward the more appropriate goal of *universal participation* (Bolt, Beranek, and Newman, 1993) on the information superhighway by the nation's students and teachers.

The universal service goal, which borrows an analogy from telephone service, means that governments use regulation to require private companies with regional monopolies to provide the public with access to minimal services at affordable prices. Still, after 100 years of telephone service and over 60 years of regulation, there are few telephones in schools today. Few school districts in the country have seen the educational and communication services on the telephone network that would justify both the ongoing service costs and the up-front investment in a local school-based telephone infrastructure that would ensure universal participation by students and teachers. Many more factors than access will be needed to justify an investment in a computer-based telecommunications infrastructure that provides the pathway to universal participation on the information superhighway.

The national debate on education today stresses as its goals not just access to education but instead high standards of what students know and can do. It is the active participation on the information superhighway that helps students develop the planning, interpersonal, informational, technological, and communication skills required by the knowledge-based citizens and workers of the 21st century. If such skills are the goal of long-term federal policy for K-12 education, then universal participation is the appropriate strategic goal of federal policy.

Neither federal nor state government action can assure K-12 education both universal access to and widespread student utilization of the information superhighway. Smart regulation and investment can, however, encourage organizational changes in schools and the emergence of new educational service providers on the information superhighway. These new services will, in turn, provide the incentive for local communities to invest in the development of an adequate and sufficient local school and school district telecommunications infrastructure.

Government goals, at both the federal and state levels, must be to ensure K-12 access to the future national information infrastructure (NII), to provide equity for all students, and to support the development of information resources for education. Such goals, however, will not be realized unless public and private initiatives combine to enable the development of a real economy of schools and learning enterprises on the information superhighway.

K-12 education is a totally different sector, economically, than banking, health, or libraries. Though educators today know well the effective learning activities that can take place on the information highways, there remain many roadblocks to significant K-12 traffic. K-12 is still very much a cottage industry of some 100,000 disparate school units, not easily subject to economic integration or rationalization. While banking, health care, libraries, and higher education have all invested substantially in data networks, K-12 lacks a telecommunications infrastructure at all levels—national, state, school district, and, most important, at the school site. While other economic sectors provide obvious markets that the NII will facilitate, such as money transfer, medical information and report sharing, information resources, research, and video on demand, K-12 opportunities will not emerge without structural changes in school organization and the parallel development of new educational enterprises that utilize the new telecommunications potential.

What will be the triggering event for schools and school districts to reallocate current expenditures and invest new dollars to equip their students and teachers with the vehicles, training, local roads, and on-line ramps to the superhighway? Can government investment and regulatory activity be designed to spur the organizational changes in schooling and encourage the growth of new educational service providers that, in turn, can combine to exploit the educational potential of the future information superhighway?

NO MYSTERY AS TO THE EFFECTIVE APPLICATIONS OF NEW COMMUNICATIONS TECHNOLOGIES

There is no mystery today among educators as to what the effective and powerful applications of communications technology in education are. Using technologies such as computers, CD-ROMs, videodiscs, phones, cable, broadcast, satellite, local area networks, and wide-area "internetworking," students and teachers today in exemplary technology-using schools can do their work, access information, communicate via electronic mail with each other and with mentors, engage in professional collaboration and student collaborative project work, go on electronic field trips, create virtual learning communities, and receive and use course and minicourses from any number of educational service providers. Some exemplary uses include the following:

- In July 1990 a teacher named Nikolai Yakolev got on a train with five other teachers and 20 students in Moscow and traveled 25 hours north to Kem, a town at the rim of the Arctic Circle. They went there because that was the best place on earth that year to view the solar eclipse. There, with only some remnants of the Red Army and some black flies as companions, they established a scientific plot and collected data before, during, and after the eclipse on the behavior of the animals and plants and the changes in light and temperature. They collected this data with probes connected to a portable personal computer (PC) and used Kem's only phone line to transmit the data to a computer at the Technical Education Research Centers (TERC) in Cambridge, Massachusetts, where it was stored and made available for schools throughout the world to compare with data from the forthcoming 1991 solar eclipse in Hawaii.

This project, conducted by the Moscow-based Institute of New Technologies (INT), was a forerunner to the Global Lab Project. Boris Berenfeld, INT's associate director, subsequently came to the United States to work at TERC, where he now directs the Global Lab Project, a project highlighted recently in the Clinton administration's report, *The National Information Infrastructure: Agenda for Action* (IITF, 1993). Today, students in the Global Lab Project's 101 schools in 27 states and 17 countries establish environmental monitoring stations to study climate change, monitor pollutants, and measure radiation. Schools then share the data over the network with each other and with scientists to gain a global perspective on environmental problems.

• Students at McKinley High School in Hawaii gave a presentation in Japanese to a real audience of students in Kanazawa, Japan, in real time. After they finished with this hour-long presentation, the class in Japan likewise gave its presentation back in English. Both presentations were transmitted inexpensively by video telephones through a project called Teleclass International. Students at each site prepared for their presentations for a few months in advance. The fact that they were able to link up with their peers in real time and see their faces made an enormous difference in the whole learning effort.

• A group of 25 teachers met in Indiana in the summer of 1993 for a week-long school restructuring institute. They were researching exemplary nationwide school restructuring efforts and developing a resource report. Using a "Pic-Tel" videoconferencing system, they held daily two-hour seminars with school restructuring experts in Massachusetts, Rhode Island, California, and elsewhere. The external experts were able to talk with them and to transmit video showing scenes of innovative schools throughout the country.

• Students in the Midwest have worked on scientific projects and have communicated via audio, video, data, and pictures over the Internet with scientists at various universities and in industry. The project, COVIS or Collaborative Visualization, uses scientific visualization software to enable students and consulting scientists to share and display their data and, at the same time, in a video window, engage in a real-time, two-way active discussion.

• Via satellite links in museums, schools, and universities and through cable, over 1.5 million students and teachers from around the world have participated in real science through interactive video and audio links with scientists from the Jason project. They have observed scientific expeditions and communicated with practicing scientists, in real time, in the Mediterranean Sea and at the Galapagos Islands, the bottom of Lake Ontario, the site of the wreck of the Titanic, and the coast of Baja California. The Jason V Classroom Network planned to air Expedition Planet Earth from Belize via satellite and the Mind Extension University cable network in the spring of 1994.

• Students throughout the country watch daily 12- to 15-minute news programs through CNN Newsroom and Whittle's Channel One program. CNN Newsroom schools are able to receive a daily textual curriculum guide via any kind of telecommunications service or via the cable line. In addition, the cable industry's Cable in the Classroom organization provides educational use of cable programs throughout the United States. Over that same cable line, using a special modem, students are able to receive an Associated Press-like service called Xpress Xchange, which gives them news, in a steady data stream, from all the major news agencies of the world. They can capture whatever stories they preselect by subject into their computer and save them to disk. To facilitate a debate about what is going on in Somalia, for example, they would collect stories from AP, Reuters, TASS, Xinhua, Agence France Presse, and so forth. Then teams from the school would take different national perspectives in the debate.

• Students take satellite and cable courses in foreign languages, mathematics and science, and the humanities. Distance learning, usually with one-way video and two-way audio but sometimes also with two-way video, provides courses not available to schools because of geographical isolation or economic limitation (OTA, 1989). Many of the organizations delivering these courses are Star School grantees, such as the Satellite Educational Resources Consortium, TI-IN United Star Market, and the Massachusetts Corporation for Educational Telecommunications (MCET). MCET also delivers curriculum modules that support teaching via satellite, Picture Tel, videodisk, and videotape.

• Students in Fairfax County, Virginia, have taken electronic field trips to Germany since the fall of the Berlin Wall, to China, to Wales, to the countries of the former Soviet Union, and to NASA (National Aeronautics and Space Administration) space centers. These live, interactive visits use video and satellite communications, telephones, computers, and cable for local distribution. Fairfax has shared these visits with schools around the country.

These new learning activities and educational services on the information superhighway have expanded significantly over the past several years. Over 1.5 million students around the world have participated in the Jason project. Over 14,000 schools watch the daily 15-minute news program, CNN Newsroom, and 10,000 schools receive Whittle's Channel One. Together, these news programs reach nearly 25 percent of U.S. schools. According to the U.S. Department of Education, distance learning is now used in 6,000 schools for 20,000 elementary and 20,000 high school students. The TI-IN network, transmitting video via satellite from Texas, reaches 1,100 schools in 40 states (Bruder, 1991). MCET reaches 37 states with its curriculum modules.

These applications, whether carried on phone lines, cable lines, broadcasts, satellite, or through the Internet, are the wonderful side of these communications technologies. The unfortunate characteristic, however, that unites these examples is that few have demonstrated independent commercial viability. Customers for these services include schools and state agencies. Most of the schools that receive these services pay a user fee that is not sufficient to cover costs. States, as the governmental unit responsible for schools, add support, as do foundations, and federal government agencies like the National Science Foundation or the U.S. Department of Education.

The emerging industry of educational service providers is frail and fragmented. The real market, from schools, school districts, and states, is too limited today to support the range and quality of services that schools need. The Midlands Consortium, one of the early Star Schools grantees, shut down after the Star Schools money ended. Schools are not paying customers for the daily news programs. They receive the CNN Newsroom program, which is subsidized by Turner Broadcasting, by paying minimal or no cable use fees and the Channel One news program by agreeing to have students watch two minutes of advertising. There are other factors, however, including the current organization of schools and the lack of a telecommunications infrastructure at school sites and school districts, that impede these developments. State and federal support will, of course, continue to be necessary to encourage this emerging industry, but we are going to have to look at the K-12 industry a lot closer to see how market niches might emerge to foster the development of the new educational service providers.

SNAGS, BARRIERS, AND ROADBLOCKS ON THE INFORMATION SUPERHIGHWAY

Imagine what the information superhighway looks like to a teacher in one of our 100,000 K-12 schools in the United States. First, most teachers don't even know about it. Few have external phone lines and most school district business offices will not approve the open-ended purchase orders needed for phone service. Some teachers have phone lines but have outmoded Apple II-E's or early IBM PCs as workstations, with interfaces that cannot support the newer software for easy navigation on the Internet. Some teachers have a phone line, a computer with an effective interface, a modem, and a connection, through a local university, to the Internet. But while they can communicate with colleagues in Moscow, they can do little with their primarily unconnected colleagues and students and the parents of the students in their own district. Few teachers in the country use computers that are on local area networks (LANs) connected to wide area networks (WANs), as is common in higher education, business, and the research community.

Market Data Retrieval reports that 49 percent of school buildings have LANs. Most of these, however, are relegated to a single lab for limited instructional purposes, such as drill and practice or remediation. Occasionally, an entire building is wired for administrative purposes, including the reporting of attendance data and grades. While a survey by the California Technology Project found that 10.6 percent of 400 schools surveyed had LANs connected by modem to WANs, further research showed these connections were for administrative purposes and that there were "no connections between instructional LANs and instructional uses of wide area networks" (Newman et al., 1992).

Typically, schools and group of teachers who might want to develop the local infrastructure for school networking do not have budget control at the school site level. They have difficulty paying for the phone lines. The barrier to putting the nation's 100,000 schools on to the Internet is not the capacity of the Internet, but the lack of local school and school district infrastructure and the difficulty in any locality of financing the end-user equipment and the LANs that represent the vehicles and local access roads to the information superhighway.

Many states are currently developing plans to give all schools access to the Internet. In most cases this means dial-in access by single computers and modems and use of data but not video communication. Some states, such as Iowa, plan a more extensive fiber optic backbone with county points of presence to which local schools can connect, permitting both data and video communications. This is the kind of infrastructure states should build. This still leaves, however, two tasks for local school districts. First, districts will have to run the fiber to the "curb" of the local school and, second, wire the school and build the in-school voice, data, and video distribution system. For all members of a school community to "drive" on the information superhighway, schools will need their own local telecommunications infrastructure, including an Internet server and router, cable, and satellite connections, and internal voice, data, and video distribution, all of which requires significant investment. The most extensive wiring of schools in the country, by Whittle's Channel One news program, has, unfortunately, done little to meet this need. Channel One schools get a fixed satellite dish, receiving equipment, 19-inch monitors in every room, and a limited video distribution network, with only the ability to send out two signals to classrooms and only receive from one satellite. It is much too constraining, not the real video distribution that schools need, nor does it do anything to set up a school-wide data communications network.

The costs of providing real access to all U.S. students on the future NII are significant. Educational researcher Henry Jay Becker estimates the annual personnel, hardware, and software costs at nearly $2,000 per student for developing expertise in technology use among teachers and providing students with a learning environment characterized by project-based learning, gathering information from diverse sources, and electronic communication with "students all over the world, with scientists engaged in real world inquiry, and with databases of enormous magnitude" (Becker, 1993).

That is nearly one-third the current annual cost per student in most U.S. school districts and would amount to approximately $90 billion in additional costs annually for all the nation's schools. Such an investment is unlikely to happen, except in wealthy districts or in schools and districts where there is a clear understanding that the up-front investment will yield real and rapid dividends, such as better and more appropriate student outcomes and economies in the costs of schooling.

The Telecommunications Industry and K-12 Education

In the real economy, building the home-school, school-community, and wide-area connections will require partnerships between the phone companies, cable companies, the schools, and the states to plan the local and regional infrastructures.

Cable shows how the real economy works. The spread of cable to U.S. schools has little to do with the schools as a current market for cable products. Instead, schools are the beneficiaries of a struggle for public policy leverage and regulatory advantage between the cable industry, the regional Bell operating companies, and the other telephone and telecommunication companies (telcos). Displays of public spirit and partnerships are the currency in this battle. To its credit, the cable industry competes by providing excellent educational programming to schools and, through Cable in the Classroom, an industry organization, obtains and identifies program rights for recording and use in schools. Cable in the Classroom and CNN have also done a wonderful job of encouraging regional and local cable multiple service operators to wire schools and their classrooms, not just run a cable drop to the school door, as was the practice until recently.

In the distance learning sector, there are many good examples of programming services that are encouraged by the federal government's Star Schools initiatives, but in most cases none of these are real commercial successes. Federal and state support is critical to nurture the development of these service providers.

By providing video equipment to schools in exchange for students watching two minutes a day of commercial time, Whittle's Channel One program tapped an economic niche by giving advertisers access to a well-defined teenager audience. For this service, advertisers pay $157,000 a half-minute to reach 8 million students, yielding Channel One a daily gross of $628,000 (Kubey, 1993). This has been very controversial in school districts and states all over the country, but it did demonstrate that there was a small niche in the marketplace. It also shows the limit of that niche, as there is hardly any room in the school schedule to expand that kind of advertiser-supported educational programming into schools.

Many companies are now eyeing the potential of the NII as the delivery system for future electronic and video educational services such as customized curriculum, thematic units, customized textbooks, courses, modules, and electronic field trips. Some of these companies come out of the educational technology sector, like the Educational Management Group, but most represent new alliances from companies in the publishing, printing, cable, and telecommunications sectors.

The Organization of Schooling

> "Informating" is organizing to exploit information technology.
> —Shoshana Zuboff, *In the Age of the Smart Machine* (1988)

Besides the substantial cost of building the local school telecommunications infrastructure, there is also the problem of the organization of schooling. Schools are not really organized to exploit this new world. Schools and school districts need to change profoundly in order to exploit these new technologies, to "informate." For students of today to become the citizen knowledge workers of tomorrow's global economy, and for students to acquire the high-performance skills set forth in the SCANS[1] report, *What Work Requires of Schools* (SCANS, 1991), schools will have to change to have, in a sense, 24-hour school days and 365-day years, operating around the clock by telecommunications.

Education in what James Mecklenburger calls a global village school (GVS) would be appropriate to both today's and tomorrow's world. According to Mecklenburger, "A GVS will be concerned round-the-clock for learning. A school building will not define the school; a building may be the 'headquarters' of a community of learners but not the sole location. Electronically, through voice mail and various networking schemes, there are many ways even today that home, school, and other institutions near and far can be intimately connected . . ." (Mecklenburger, 1993). Changing schooling requires the changing of all elements of school practice and organization simultaneously, including "school organization, curriculum, assessment, technology, and the learning environments" (Pearlman, 1993).

Students have to be the real workers in education, the ones who actually produce and do the work. There is a very good vision of how schools might be organized in British management expert Charles Handy's book, *The Age of Unreason* (1989). Handy describes how the "shamrock organization," or what others are calling a virtual corporation, would operate in education. It should be stressed that "virtual corporations" are not synonymous with "virtual reality"; instead, they are real corporations that are becoming leaner, more engaged in core activity and are outsourcing much of their work to other companies or to consultants. It is in this sense that the "virtual school" may emerge, functioning as Handy says, "as a shamrock with a core activity and everything else contracted out or done part time by a flexible labor force. The core activity would be primarily one of educational manager, devising an appropriate educational program for each child and arranging for its delivery." Much of that delivery could be on the future NII.

K-12 education is in much the same place today as the service industry was in the mid-1980s. Despite some of the great capabilities available in schools today (including single desktop machines with associated communication links), critical mass is nowhere in sight. Schools and school districts in the United States are spotted with technology-using teacher pioneers, but widespread utilization and its associated productivity gains remain distant and elusive. The school site communications infrastructure is not in place and, even if it were, schools are not organized to exploit it.

If schools are to change in the direction of shamrock organizations, they will need to outsource those learning activities not provided at the school site. This means that new educational service providers will have to arise locally, regionally, nationally, and internationally to deliver courses, minicourses, project activities, curriculum, materials, and apprenticeships to meet the range of learning needs and learning styles in the shamrock school. Enabling this new industry of education providers is, of course, the potential of the information superhighway, or NII.

K-12 Education and the Real Economy

Nobody heard the story on the news wires that the Los Angeles Unified School System bought out the San Francisco Unified School District. No one heard about the Whittle bid for majority control of the New York City Public Schools. Nor the merger of the Washington, D.C. Public Schools with those of Fairfax County, Virginia, and Montgomery County, Maryland. But everyone has heard about the rash of mergers in the health industry, in banking, in telecommunications, and in many other sectors of the global economy. No one heard about any of these kinds of economic developments in K-12 education because it is a distinct kind of economic and organizational entity, not easily subject to economic rationalization. While schools are governed by local, state, and national entities, in no sense is any of these a system in the way that banking enterprises are integral national or global economic units. Chris Whittle's stalled efforts to establish a national network of Edison schools did aspire to such systemic economic and organizational integration. Despite its failure to raise sufficient investment funds, the Edison project does raise the proper question of how, whether private or public, to put together a network of schools, whether in one location or around the country, that actually work together and produce some savings by sharing curriculum, programming, and teachers through a national delivery system of educational services.

The Clinton administration's NII initiative aims to loosen the regulatory noose around phone and cable companies in order to get the private sector to build the information superhighway, yet retain regulation as a way to ensure a "public right of way on the information superhighway" for schools, libraries, and museums (*New York Times*, 1993).

Schools, of course, will need much more than a "public right of way." "Universal access" to phone service has not led to phones in schools nor to significant investment in teacher and student workstations, school and district LANs, Internet nodes, and training among K-12 schools and school districts, as it has in higher education and in the research community. The "public right of way" to the information superhighway only deals with a portion of the costs. What will be the triggering event for schools, school districts, and local communities to reallocate current expenditures and invest new dollars to equip their students and teachers with the vehicles, training, local roads, and on-line ramps to the superhighway?

GETTING TO CRITICAL MASS:
BUILDING THE 21ST-CENTURY INFRASTRUCTURE FOR SCHOOLING

The first person with a telephone gets little productive use if his or her neighbors and family are without phones. The first person with a car in a remote region doesn't get much local benefit until the neighbors also acquire them and local society and government build roads for the local infrastructure. So, too, in schools, things don't take off until most teachers and most children are using technology to do their work, when they can communicate with each other and with parents and community-based mentors as easily as pioneer technology-using teachers and students can today share data and reflections with counterparts in Moscow on the Global Lab project. Schools today lack the critical mass of skilled information highway "drivers" that will lead to organizational changes in schooling and to a new industry of educational service providers. Besides a lack of physical infrastructure in schools, school districts, and states in terms of local-area and wide-area networks, another serious problem is that schools today are filled with outmoded technology equipment. More than half of the computers in schools today are five or more years old. According to Quality Education Data (QED), less than 30 percent of the computer equipment in schools can support graphical user interfaces (GUIs) such as Windows or Macintosh. The GUI is critical to today's applications, whether in the creation of multimedia documents or in "internetworking" around the globe.

To realize schools for the 21st century requires that state and federal governments develop the policies and investment that will spur local school and school district investment in local school network infrastructure and that will assist schools in the process of envisioning, reorganizing, and redesigning themselves. Local communities will not make the enormous local investment in a school telecommunications infrastructure unless there is a clear public understanding of its perceived benefits. Despite the acquisition of some 4 million microcomputers in U.S. schools, technology has not resulted in many economies in the cost of schooling. In K-12 education about 80 percent of school budgets have to do with personnel, mostly teachers. Some private companies that are making a business of managing schools are realizing economies on custodial or food service personnel, but significant economies in schooling would require reductions in teacher personnel. This has not happened as yet, even in technologically rich schools.

The only way to realize that is by totally reorganizing schooling and making a significant up-front investment in communications technology. What would it take for schools to work effectively with fewer teachers? Students would have to be able to work much more on their own, with teacher advisors managing their education with the support of external mentors and service providers, as in Handy's shamrock organization. To support this, schools would have to establish up front a communications infrastructure with a local area network in the school connected to the

Internet and with connections throughout a wired local community. With that kind of structure in place, it is possible to imagine that significant economies could occur in school personnel. Schools today, however, face the dilemma of investing up front significant funds and incurring ongoing costs for a communications infrastructure before there exists the availability of sufficient educational services on the information highway that are cost efficient and effective.

This, of course, is the critical mass, or chicken and egg, problem. Without widespread local school infrastructures, no one will invest in creating new educational service companies. Without widespread and cost-efficient educational services, local school districts won't invest in local school infrastructures. To solve the critical mass problem requires the progressive and simultaneous development of:

- The information superhighway and the "public right of way" (highway and on-ramps);
- A new industry of educational service providers that deliver distance learning, curriculum, educational resources, project-based learning activities, and so forth (program);
- A reorganized system of schooling and schools (organization); and
- The local school telecommunications infrastructure (local roads and vehicles).

There are many attempts today to reinvent and redesign the American school. These include efforts by school districts to build new kinds of schools (Pearlman, 1993), the efforts of the design teams sponsored by the New American School Development Corporation, schools launched by Charter School legislation in many states, and private efforts, like the Edison project. All of these attempts focus on the question of how to develop schools that have both better student outcomes and outcomes more appropriate for the 21st century; that are more economical and are each tied to a system of schools, whether in one location or around the country; that actually work together and produce some savings by sharing curriculum, resources, teachers, project activities; and so forth.

Today, K-12 education, despite the governance of school districts and state systems, is effectively a cottage industry with 51 state units, 15,000 school district units, and 100,000 school units that exhibit little or no economic integration. These school districts and state systems will have to evolve in a way that goes far beyond their current governance and regulatory role to use the new information infrastructure to provide services and bring about efficiencies in the way that a global corporation like Citibank or a large health maintenance organization utilizes networks to rationalize and economize on the delivery of services.

As a nation, we will need many experiments on the new design of schooling and the utilization of telecommunications network services over the next several years in order to understand how to exploit its potential. These experiments will come from both government investment and private and public partnerships.

Despite the fact that the current Internet owes much of its development to prior governmental support, the NII of the near future will be largely a private venture and will swamp the current Internet in size and power. Universal digital access and the "public right of way" for schools, libraries, and museums will come about less through government investment than through private development.

The key sectors of the telecommunications industry—cable and the telcos—are actively engaged with school partnerships in their contest for public policy leverage. Smart schools, school districts, and states will use these partnerships to show what can be done effectively and productively in K-12 education with an advanced communications infrastructure.

GOVERNMENT INVESTMENT AND REGULATION
TOOLS TO GET K-12 TO CRITICAL MASS

Government has a key role in enabling the development of an information superhighway that fosters the simultaneous development of new school organizations and the educational service providers to serve them. But federal or state investment can never be of the scale to provide nationally to local schools and school districts the necessary level of school site workstations and LANs, training, and nodes to make them full partners on the information superhighway.

Federal and state policy initiatives should be aimed at overcoming the critical mass problem so that local schools and school districts see the benefits of investing in the local infrastructure. Today, few classrooms have phones. Tomorrow's schools will have digital links not just between Denver and Moscow but also, and more importantly, between the local school and its own community, parents, students, and teachers, so that students will be able to communicate with teachers, parents, community mentors, and international mentors and resources and share messages, ideas, data, and multimedia student work. To raise the funds needed to build the local infrastructure, local schools and school districts will have to understand the benefits of these information-age skills and activities and be able to persuade the local citizenry to invest, through special bonds and appropriations, in such a capability.

Federal and state government policy can support these developments. Government can be a customer of new services and can invest directly in, and promote through grants, the following:

- Public educational information resources;
- Long-range planning by states, school districts, and schools;
- Staff development;
- Software and interface development for internetworking;
- Project-based learning modules;
- New educational enterprises; and
- Research on the effectiveness of the new learning activities and enterprises.

The federal government can increase existing support through direct NSF or National Telecommunications and Information Administration grants, through defense conversion funds or the High Performance Computing and Communications Initiative, the Star Schools program, and the U.S. Department of Education. In the regulation sector, tariffs could be regulated and targeted subsidies established to enable school consumers less costly access to the information highway. The state of Georgia gives schools the same below-tariff rate as state agencies and, together with a less than 1-cent surcharge, has generated $35 million annually for an education technology trust fund (Kessler, 1993).

Government must respect that the most significant investment will come from the private sector, when private interests coincide with the public interest, and not because private sector subsidies are specifically required by regulatory rules. Linkage requirements to builders of the information superhighway to give schools a "public right of way" can, however, be used smartly to foster win-win public-private partnerships between schools and the telephone or cable companies. K-12 education also needs to develop partnerships with other public service sectors, including libraries, museums, social services, and health care; to press for government action; and to lobby effectively with the telecommunications and cable companies.

Building the information superhighway, the Internet, and the future digital NII is not enough. That is just infrastructure. The key development, besides national, state, and local infrastructure, is the availability of quality educational programs and services.

Few technologies have made any impact on the organization of K-12 education, whether radio, broadcast TV, satellite TV, cable TV, or computers. Accompanying these developments, however, have been an increasing number and quality of both public and private, but free, programming services. With the development of the NII and "shamrock schools," one can envision the parallel development of local, regional, and national program providers that are part of the real economy, that is, neither public supported nor advertiser supported but instead paid for through service fees by schools that have been able to reorganize and reallocate their expenditures to pay for such services. These services would include direct instruction courses, minicourses and modules, thematic blocks for project-based learning, and an array of curriculum materials and information resources for school and home use. They could also provide the basis for companies that manage networks of public or charter schools dispersed throughout the country.

A Seat at the Table

Like other economic sectors, K-12 education is faced with a critical mass problem. While commercial viability of NII-based services is evident in such sectors as banking, financial services, and new video-on-demand services, K-12 education will require substantial state and federal support to nurture the development of school-site telecommunications infrastructures and new educational services. K-12 education needs a seat at the table in national efforts to develop the NII. To be effective, K-12 education will have to get its voice together in a better way as a sector and in coalition with other public service sectors. The federal government could aid that process by financing national teleconferences for the education sector to learn the issues of the NII and build its collective voice.

State and federal governments are critical players in promoting universal access to and widespread student utilization of the future NII. Through investment and regulation, states and the federal government can promote the new school organizations, the new educational enterprises, and the local school-site telecommunications infrastructures that will make the information superhighway a roadway for today's and tomorrow's students.

REFERENCES

Becker, Henry Jay. 1993. "A Truly Empowering Technology-rich Education—How Much Will It Cost?" *Educational IRM Quarterly* (Fall).

Bolt, Beranek, and Newman Inc. 1993. *Getting the NII to School: A Roadmap to Universal Participation*. Bolt, Beranek, and Newman, Cambridge, Mass., December.

Bruder, Isabelle. 1991. "A Guide to Distance Learning," *Electronic Learning* (November/December).

Handy, Charles. 1989. *The Age of Unreason*. Harvard University Press, Cambridge, Mass.

Information Infrastructure Task Force (IITF). 1993. *The National Information Infrastructure: Agenda for Action*. Information Infrastructure Task Force, Washington, D.C., September 15.

Kessler, Glenn. 1993. "Treat Schools as State Agencies for Telephone Rates," *Inventing Tomorrow's Schools* (May/June).

Kubey, Robert. 1993. "Whittling the School Day Away," *Education Week* (December 1).

Mecklenburger, James A. 1993. "To Start a Dialogue: The Next Generation of America's Schools," *Phi Kappa Phi Journal* 73(4):42-45.

Newman, Dennis, Susan Bernstein, and Paul A. Reese. 1992. *Local Infrastructure for School Networking: Current Models and Prospects,* BBN Report No. 7726. BBN Systems and Technologies, Cambridge, Mass., April.

New York Times. 1993. "Gore Views the Data Highway," December 22.

Office of Technology Assessment (OTA). 1989. *Linking for Learning: A New Course for Education,* OTA-SET-439. U.S. Government Printing Office, Washington, D.C., November.

Pearlman, Robert. 1993. "Designing the New American Schools," *Communications of the ACM* 36(5):46-49.

Secretary's Commission on Achieving Necessary Skills (SCANS). 1991. *What Work Requires of Schools.* U.S. Department of Labor, Washington, D.C.

Zuboff, Shoshana. 1988. *In the Age of the Smart Machine: The Future of Work and Power.* Basic Books, New York.

NOTE

1. SCANS (1991) defines a framework that can serve as the foundation for the establishment of education goals and objectives in future U.S. citizens. Whether they work on a shop floor or in an executive suite, they will need to master the following competencies and foundations skills:

COMPETENCIES:
1. Resources: identifies, organizes, plans, and allocates resources.
2. Interpersonal: works with others.
3. Information: acquires and uses information.
4. Systems: understands complex interrelationships.
5. Technology: works with a variety of technologies.

A THREE-PART FOUNDATION:
• Basic skills: reads, writes, performs arithmetic and mathematical operations, listens, and speaks.
• Thinking skills: thinks creatively, makes decisions, solves problems, visualizes, knows how to learn, and reasons.
• Personal qualities: displays responsibility, self-esteem, sociability, self-management, integrity, and honesty.

Future Roles of Libraries in Citizen Access to Information Resources Through the National Information Infrastructure

Clifford A. Lynch

INTRODUCTION

This paper critically examines several popular assumptions about the national information infrastructure (NII) vision that I believe are not entirely consistent with the evidence available to date and current trends. These assumptions include the following:

• Ensuring universal access to the NII and the resulting benefits are largely synonymous with ensuring universal connectivity, as has been the case with the electrical power grid or the telephone system. If such universal connectivity can be provided, immediate benefits in terms of improved access to information resources and services will follow.

• The NII will greatly increase the public's (free) access to information. The public will access digital libraries that will supplant the functions of traditional (physical) libraries. The digital libraries on the NII will be enormous storehouses of information that will include (but not be limited to) much of the existing literature of the world. Forward-looking traditional libraries are already transforming themselves into such electronic storehouses.

• In the new environment of the NII, libraries will continue (and expand) their current role as the key providers of information to the public (regardless of economic status). Libraries will be the mechanism through which we will balance the increased commodization of information with the public's need to have access to information.

• Libraries will benefit greatly from the NII. The new technology will facilitate efficiencies in resource sharing that will help to alleviate the budget crises facing libraries of all types and will improve the quality of library service nationwide, in part by reducing geographically based inequities among libraries.

UNIVERSAL ACCESS IN THE NETWORKED INFORMATION CONTEXT

The NII enterprise has two components. One component is the upgrading of the existing national telecommunications infrastructure to incorporate very high speed computer-communications facilities that will efficiently (and hopefully cheaply) move large amounts of digital information, including digital video. This is the realm of integrated services data network, asynchronous transfer mode, fiber to the curb and the home, "video dial-tone," and similar technologies. The issues are numerous and are well covered elsewhere in this volume; they include such basic public policy concerns as how to encourage the rapid deployment of the infrastructure on a broad basis, how ubiquitous the new infrastructure must be, and regulatory issues related to competition and monopoly.

The second component of the NII initiative is the set of applications that will be enabled by this upgraded telecommunications infrastructure. These applications must drive and justify the initiative: the new applications will engage and excite the public to support the effort. Indeed, the initiative has been justified with visions of a citizenry informed and empowered by greatly improved access to information and students—be they children in rural areas or adults engaged in lifelong learning activities—exploiting ubiquitous access to the world's literature to facilitate their learning. These visions have been eloquently sketched not only by the research and education community but also by political leaders such as Vice President Gore. They are indeed worthy goals for public policy, and, technically, the visions are challenging but not unrealistic. From social, legal, political, and economic perspectives, however, they are much more problematic. The difficulties are not widely understood and have not been well addressed.

Curiously, services such as person-to-person electronic mail—which are quite similar to the services offered by voice telephony, and for which we do have some understanding of demand, usage, and economic basis—are not often cited as justifications for moving forward with development of the NII, either by analogy to current phone service as a universal citizen right or as an example of a clearly viable commercial service that could be enabled by a ubiquitous NII. Perhaps they are not enough: not sufficiently compelling in their impact, not exciting enough as a potential market-place, not challenging enough in terms of the base telecommunications technology necessary to support them.

The new telecommunications infrastructure will enable other information-content-intensive applications that may create entire new business sectors. These are more compelling to the corporate world and are not often mentioned as public policy justifications—in particular, the creation of a marketing paradise on a previously unimagined scale. Such enterprises will succeed or fail based primarily on the marketplace; their success or failure is not a public policy issue. My opinion, which is somewhat at odds with the view represented by the enormous investments currently being made in the alliances and acquisitions involving regional Bell operating companies, media companies, and cable television companies, is that the marketplace viability of many of these commercial services (and the time frame in which they will become profitable) is still uncertain.

In framing access to information services and resources through the NII as a public policy issue, we are fundamentally in conflict about what we are trying to accomplish. We welcome the economic growth that sales of information access through the new networks are likely to represent. For those who can pay, it seems clear that the NII can only increase the range of information that is accessible for purchase, since it will add to all of the current sources of information, and the NII forms a hospitable environment for a wide range of fundamentally new types of information content and information services.

But as a society we have beliefs about the rights of citizens to have access to a wide range of information and the importance of such access in contexts like education. The NII is particularly appealing because it offers a technological environment that can expand such public access to information to support the visions of greatly improved education discussed earlier and can facilitate equal access to knowledge by citizens in rural as well as urban areas. In today's world we welcome and support the diversity represented by libraries, bookstores, broadcast television and radio, and printed newspapers and magazines sold both at newsstands and by subscription. While all of these enterprises compete with each other to a limited extent, they have come to coexist fairly comfortably (though with continuing minor conflicts around the boundaries). In a new world of widespread distribution of electronic information through networks (under business terms we can only guess at today), it is unclear that all these institutions will continue to maintain independent existences. They may converge and coalesce in complex, competitive ways. With the exception of libraries, all of the other information providers and distributors mentioned above are basically profit-oriented enterprises; even the exceptions, such as scholarly societies, compete with public and research libraries, which usually are publicly supported nonprofit organizations. The trajectory of competition and convergence among the for-profit enterprises is primarily a business question (though not without some public policy implications in terms of, for example, ensuring diversity). The future role of libraries in the NII environment may represent a conflict of public- and private-sector interests. The stakes in this conflict are potentially high, not only in impact on private-sector revenues but in terms of our future society.

The conflict is not just over direct revenue—it is over the library's purchasing of information and giving it to its user community, thus depriving the information provider of the opportunity to collect revenue from sales of the information. The library is also preventing the information provider from establishing a relationship with each purchaser, which can facilitate the collection of marketing and demographic data (valuable in its own right). The library continues to provide its patrons with privacy through anonymity in a marketplace that is increasingly moving toward the massive collection of data about the behavior and interests of customers.

Beyond the admittedly compelling visions of public good that are being presented to justify the NII program, our government seems to be establishing some fundamental policy objectives for its development, such as very broad access to this new infrastructure. The need for "universal" access to redress the increasing gap between the information "haves" and "have-nots" has become an issue. To some extent this universal service objective makes explicit the universal service goals that have been evolving for voice telephone service over the past few decades and extends the entitlement for basic telephone service to an entitlement not only to NII connectivity but also to some poorly defined set of information services on the future network. The discussion of policies and implementation strategies for the NII to address the universal service objective has really focused on how to achieve large-scale connectivity; it has not addressed what, if anything, this connectivity will provide to the user other than an opportunity to access whatever services may be generally available on the NII, on whatever economic terms these services may be offered (which may include free services supported by advertising and some set of free or very low cost "public access" services).

The NII is something completely new. In our quest to gain insight into the development of this infrastructure and to continue to inform our policymaking and planning, we have studied historical analogies. These include the telephone system, the electric power grid, and various transportation systems such as highways and railroads. These analogies are often confusing and misleading because they fail to capture the essential character of computer-communications networks, such as the Internet, as the new medium of communication and information dissemination. If applications for the new telecommunications infrastructure are to be driven by information, one must carefully evaluate the societal and legal structures that have developed around ownership of and access to information resources. I think that access to information services and resources on the NII will quickly come to dominate issues about connectivity in the universal service debate.

Consider the implications of universal access to voice telephony. Roughly speaking, such universal access means that everyone has a phone, can receive calls, and perhaps can make local calls. Access to additional services is based on the ability of the end user to pay. Universal access does not guarantee anything other than the most limited enjoyment of the intellectual property of others. Universal access to the services that we expect of the future NII (and indeed the services we have used to justify its development) are fundamentally different and go beyond simple connectivity. They involve universal access to information resources, which (except for public domain and other freely accessible information resources) are someone's intellectual property. The widely ignored precedents here are advertiser-supported broadcasting (both television and radio), which provides the public with access to a rather limited set of information resources and artistic works under a very peculiar "contract" that requires the public to endure a certain number of minutes of advertising (or pledge drives in the case of public broadcasting) along with the viewed content.

Indeed, in the NII context we will be challenged as a society to define a base level of information resources that we believe must be available to all members of our society, regardless of the ability to pay. Access to some of these resources will probably be subsidized by advertisers; others will be made available directly by local, state, and federal governments; still others will be provided by a range of nonprofit groups, including, perhaps, some future "Corporation for Public Network Information" and its local "Public Network Information" affiliates. Access to those resources not provided through other means will have to be subsidized by government, just as it subsidizes public libraries to provide access to a good deal of information not available through other mechanisms.

LIBRARY ROLES IN UNIVERSAL ACCESS TO ELECTRONIC INFORMATION

Libraries defend intellectual freedom, the right to privacy in accessing information, and the public's right to have access to a wide range of information and viewpoints. Over the past century libraries have been effective advocates and defenders of these values. Today libraries successfully provide access to a very broad and diverse base of information. Individual units of information such as books are fairly inexpensive, and a wide range of materials can be acquired for a library's collection. The interlibrary loan system essentially permits a library to offer its patrons a range of information that goes far beyond locally held materials, which greatly expands the scope of materials available to every library's users. Ensuring unlimited public access to a given electronic information resource, on the other hand, appears to be a very expensive proposition, and for reasons to be described later in this paper, it appears that such electronic resources will not be able to be shared among libraries through interlibrary loans. As we look toward the issues of information access on the NII, one danger is that the goal will be reformed from an assurance that the public has access to a broad range of information to a definition of specific information resources that the public must be able to access. Such subsidized access may be limited to a few selected electronic information resources in much the same manner as school systems today acquire textbooks and the government creates and develops reports on specific topics of public interest. In the future the government may develop and mount electronic information resources or contract for these resources from the private sector for public use. In the future, support of wide-ranging intellectual curiosity and open-ended exploration of diverse viewpoints may no longer be part of the publicly supported program. One of the most troubling possibilities is that the government will select a base set of information resources to be guaranteed to the public, rather than simply subsidizing some amount of access to the broad base of "commercially published" information.

To be sure, government-subsidized information sources will not be the only voices to be heard for free on the NII. This information will be complemented by a wide range of information offered by various public service organizations and indeed by any organization with a message to deliver and the funds to support the delivery of that message. (Organizations will include advertisers; the difference between advertisers and public service organizations providing information grows more and more blurred.) Further, the cost to an individual or organization offering information to the public through the NII should be much lower than the analogous costs for print distribution or dissemination of information through the broadcast media today. But a reliance on these various free sources of information does not seem to be a sufficient substitute for the broad access to published literature provided by libraries today. Libraries not only provide access to a wide range of current information but they also conserve and provide access to a cultural and historical record. Those providing "free" information through the network will neither maintain such a record nor guard its integrity, accuracy, and balance.

Ensuring access to an appropriate range of information resources and services is not the only issue, although it may be the most difficult long-term problem. Connectivity issues must be addressed as well. People will need to invest in technology to access the NII, but many will be unable or unwilling to make such investments. Libraries may be a natural place to make available the information technology the public needs to use the network, at least at some minimum level of functionality. I also anticipate the development of a number of personalized direct information distribution services that assume their users have personal "information appliances" of some type. The ability to exploit these new types of information services will be limited in a public access environment. Such an environment will support today's user-driven searching type of information access but not the emerging model that includes software agents continually working on a user's behalf to ferret out and sift information. Users in rural areas may face continued problems with network access or with the cost of that access, depending on how various policy decisions shape network deployment and economics. And training and user support issues, particularly for those who have not grown up with the technology, will be of major importance. Meaningful universal access will require a major education and outreach program. Such a program will require not only increased emphasis on information technology literacy as part of the basic K-12 curriculum but also a means of reaching the adult population.

LIBRARIES, INTELLECTUAL PROPERTY, AND ELECTRONIC INFORMATION IN A NETWORKED ENVIRONMENT

The majority of information has never been available for free use. Someone owns it. It is worth noting that even for works that are hundreds of years old and out of copyright, publishers have used a great deal of ingenuity and have gone to a great deal of trouble and expense to be in a position to claim ownership (through edited editions, translations, packaging of out-of-copyright works with newly commissioned introductions and notes, and so on). And there is a substantial body of information that is created or compiled by the government at various levels that is "owned" by the public. I have no doubt that the Internet today and the NII tomorrow will tremendously facilitate public access to these important information resources. Legislative initiatives and projects by the national libraries and federal agencies are making many of these information resources available today. But there is also a great reluctance to see the government expand its role as a creator and compiler of information very far. I would speculate that there would be little support for a government-sponsored encyclopedia to compete with the private products available today. Most of the information that the government collects, creates, or compiles is closely related to its operations or to specific legislative mandates and public policy objectives.

For most of the 20th century, libraries have provided the public with "free" access to a great deal of this privately owned information. (More precisely, the public has funded libraries; these libraries then acquire information and make it available to the public without charge.) The basic enabling mechanisms have been the copyright law and the doctrine of first sale: The library purchases a copyrighted work. While the library cannot copy it, it is free to loan it to anyone (including patrons and other libraries). The impact of the role of libraries on the revenue to rights-holders has been limited by the fact that only one person can borrow (read) a copy of a work at a time, plus the fact that material purchased by a given library is primarily accessible only to those in fairly close geographic proximity to that library. While in theory the work is available nationwide through interlibrary loan, realistically the delays and inconvenience of moving material physically from one library to another have been such that use of library materials, just like use of the libraries themselves, has had a strong geographic bias. It is only someone with a very strongly felt and typically rather specialized information need who will pursue interlibrary loan from a public library to a research library, and research library to research library interlibrary loans are typically in support of a relatively small community of scholars.

Further, the cost of travel to a distant library has historically been much greater than the cost of acquiring a given unit of information. With the exception of a few scholars who needed to use particularly rare, fragile, or valuable material that did not travel through the interlibrary loan system, people would not travel far to use material in a library. Again, this served to limit the economic effects of libraries providing access to material on the rights-holders.

We are already seeing this set of assumptions change with electronic information resources. In many cases the cost of gaining access to electronic databases is so high that users will travel to a research library that offers its user community (which typically includes walk-in patrons as well as their specific institutional user community, in the library or accessing the resources remotely) to use them rather than to pay the prices charged to the general public. We still have a very limited understanding of how this shift is really changing revenues to rights-holders; it is unclear the extent to which users are physically traveling to libraries with electronic information resources to avoid paying, as opposed to the extent to which they are making the journey to get access to a resource that they simply could not afford under any other circumstances.

Congress has also played a role in making sure that information is available to the public. The copyright laws, following the constitutional mandate that Congress "promote the progress of science and useful arts," have crafted a balance between the need to reward rights-holders for creating intellectual property and the needs of society to make use of this property. The elements of this balance include limited duration of copyrights and specific exemptions for fair use, teaching, and preservation of the nation's intellectual heritage under certain circumstances. Some of these specific provisions in the copyright law, as supported by the interpretation of the courts, recognize that there are differences between information published primarily for mass markets and information published primarily to support scholarly communications.

In an environment of networked access and electronic information this entire framework changes. Electronic information is virtually never sold; it is licensed, and its use is regulated by a contract between the rights-holder and the user that typically limits access to the information to specific uses (and perhaps even for specific purposes), to a specific, defined, limited user community and for a limited duration of time.

Two of the major factors in determining the price of a license for a given resource are the size of a library's user community and (sometimes) the maximum number of patrons that will be permitted to access the resource concurrently. Costs are high; a system like the University of California pays about $100,000 a year for the right to provide its user community with access to a large citation database or the full text of a few hundred periodicals. In many cases the electronic resources are unique offerings (and critical to scholarship in a given discipline), and there is little competition to drive prices down. Many of the public policy compromises that characterize copyright law are not usually accommodated in licenses for electronic information resources. No one knows how such compromises and the broader objective of the free flow of scholarly information can be effectively implemented in the electronic environment without unacceptable damage to the revenues of the rights-holder, barring large-scale, systemic restructuring of the scholarly publishing system. With mass market information (which includes such things as newspapers, which are also important to the scholarly community), there is even less experience and less clarity regarding terms and impact of licenses for electronic information. Very little of this mass market material has been licensed electronically to institutions such as libraries except in stand-alone formats such as CD-ROM disks.

A few other points should be stressed. Libraries cannot begin wholesale digitization programs to convert their existing print collections into electronic form; it is not permitted under existing copyright law for material that is still under copyright, and most of the most heavily used information is still under copyright. Rights-holders are under no obligation to make their information available to libraries in electronic form under any terms (regardless of whether the library already holds a print copy). Availability is much more of an issue for electronic information than it is for print: if a publisher makes a print product generally available, it is difficult to prevent a library from acquiring it under the same terms it is offered to the general public. (Differential pricing for libraries—particularly of journal subscriptions—has been imposed by publishers for a number of years; the ethics and legality of this practice remain a very vexed topic, although the practice is now widely accepted. It is hard to distinguish where differential pricing ends and a refusal to license to libraries begins; a library subscription that is 3 times the cost of an individual subscription may still be affordable, but a library subscription priced at 100 times the cost of an individual subscription likely represents what is in effect a refusal to make the material available to libraries.) In many cases, rights-holders are making a comfortable profit from print publications and perhaps from some limited direct electronic licensing to specific users (such as the business community). They have little, if any, incentive to make their information available electronically to libraries; and if they are willing to do so at all it may be at a premium of many times the cost of print. Libraries, already strapped for funds to continue any sort of acquisitions program, may be unable to afford or at least cost justify an investment in many publishers' electronic information, particularly if it is already available and affordable in print form. And the publisher has little incentive to lower the price because the print profit is already sizable. Also, publishers are being cautious about making electronic information available because pricing patterns are not well established, even for backlists or journal backfiles. They may not make retrospective files available to libraries at all, or only at very high costs. Libraries, concerned with their ability to maintain current acquisitions, will likely assign a very low priority to funding relicensing of materials in electronic formats that they already own in print or can get through interlibrary loan.

Further, any license agreements that a library may be able to negotiate typically do not accommodate the present interlibrary loan system. Information licensed by one library cannot be shared with another library's patron community. This is particularly frustrating since technically electronic information can be moved much more swiftly and economically than printed materials through an interlibrary loan system, thus promising improved service for library patrons nationwide and reduced costs for libraries that are increasingly relying on the interlibrary loan system to compensate for their dwindling local purchasing power. In some ways the interlibrary loan model does not even make sense for electronic information resources—technically, any patron at any library should be able to access resources at another library across the network without involving an interlibrary. Through the ability of networks to erase the accidental constraints of geography, anyone in the world has potential access to electronic information offered by a given library on an equally convenient basis. Electronic library collections are freed from their geographic anchors by networking technology. But facilitating such access, which clearly runs counter to a rights-holder's interests in maximizing revenue, is prevented by license. Indeed, part of the motivation to shift from sale to license has been specifically to address this issue. But although some license terms could work, to an extent, in an interlibrary loan environment (such as ensuring that only a fixed maximum number of people have access to an information resource at one time), the information providers have not been enthusiastic about experimenting with such agreements. What motivation do they have, after all?

In a networked environment libraries are being forced to be much more explicit in defining their user communities as part of their license negotiations. For a research library it is typically phrased "the faculty, students, and staff of University X"; for a public library it might be "the residents of township Y or county Z." These definitions can get cumbersome. Are affiliated hospitals included in the case of a university library; are people included who work for businesses in a city but don't reside there? There is still some recognition of public access, though this tends to be structured around geographically based traditions. For example, many libraries still propose to include as part of their user community anyone who is physically present at the library, regardless of who they are and how far they have traveled. Many rights-holders have been willing to accommodate this kind of limited public access provision. But such definitions ignore—indeed run directly counter to—many of the real advantages of the networked environment.

THREATS TO PUBLIC LIBRARIES IN THE NETWORKED ENVIRONMENT

The public libraries have always been on the front lines in making information accessible to the general public. Typically, the research library community has supported them in this effort and has also done a great deal to make information directly available to the public, though preferably through the interlibrary loan system rather than by large-scale direct service to the general public. In the new world of the NII it is natural to try to continue to keep the library community in this role and indeed to call for an expanded role particularly for public libraries in providing access to electronic information, both that provided by the government (building, for example, on the precedent established by the Federal Depository Library Program) and that made available on the network by commercial and noncommercial rights-holders and information providers.

Of all the types of libraries, however, public libraries may be in the deepest trouble. Their budgets have been slashed nationwide over the past decade, leading to massive curtailments in services. Unlike research libraries or corporate special libraries that are typically service units within their parent organizations and viewed as providing essential services to those organizations, public libraries are sometimes valued as providing a service to a community that is relatively peripheral compared to such services as public safety. Because of the very large size of their user communities, public libraries often face large costs for licensing electronic information resources. Because they tend to provide a great deal of access to mass market rather than more purely scholarly information, they are considered by some information providers as potential competitors to the information providers' efforts to market information (particularly electronic information) directly to the public, and prices for licenses are set accordingly. In order to meet the challenges of the information age they will need to make massive investments in both information technology and staff training. It is not simply a matter of ensuring that the network reaches public libraries, but also that the public libraries have a sufficient installed base of hardware, software, and sufficient competent staff to make use of these network connections as an effective way of providing access to information for their patrons.

The basic funding mechanisms for public libraries are also likely to come into serious question in the networked information environment. Historically, most public libraries have drawn their support from their geographically defined user community through various forms of taxes. In a time when any person in the world can potentially use any public library through the network, geography no longer defines the user community for a public library, and thus geographically based funding may no longer make sense as a means of supporting these libraries.

There is a great irony here. In the print world each community funded its local libraries, and these libraries served as access points to the national library collections through the interlibrary loan system. In the networked environment, collections are local to a specific library and its user community, and each library can compete equally with all other libraries on the network for patrons who would join their user community (and presumably pay for such membership as a means of funding the library). Commercial organizations are also free to appear on the network to compete with libraries for patrons. Given the opportunity to aggregate a critical mass of interested customers for information in any imaginable specialized area because of the national and international scope of the network, a wide range of newly economically viable niche commercial information providers may evolve on the network, much as such providers exist as specialty publications and shops that do most of their business by mail order. It is interesting to speculate about the potential effects of users "joining" the library or libraries of their choice through the network. Many geographic locales that currently fund libraries are quite diverse, economically, demographically, and culturally. One wonders if such diversity will continue to be found in the electronic communities that fund electronic libraries or whether we will see a much greater homogeneity among the patrons of a given "library" on the network.

Realistically, another aspect of this problem is that the economically disadvantaged or those uncomfortable with technology are least likely to own the network connections and information technology needed to access libraries over the network. They will continue to visit the library physically and to use it as a place where they can obtain not only access to information generally but also the information technology necessary to use electronic information resources on the network (which might include important community resources, such as listings of employment and educational opportunities or information about social services). The more prosperous, technologically literate people may have abandoned the geographically local public library for remote electronic libraries on the network, thus eroding the community commitment to continue to support the local library.

DIGITAL LIBRARIES AND LIBRARY SERVICES

I am not fond of the term "digital library." The term "library" is used to refer to at least three things: a collection, the building that houses that collection, and the organization responsible for it all. As organizations, libraries acquire, organize, provide access to, and preserve information. These are the primary functions of a library, though in many cases they have been augmented with more extensive responsibilities for training and teaching, for providing or facilitating access to social services, or for managing certain types of government or institutional information resources. When one considers libraries as organizations, it is clear that they deal with information in all formats—print, microforms, sound recordings, video, and electronic information resources. From this perspective, the term "digital library" doesn't make much sense and provides a very one-sided view of the mission and operation of a library.

It is certainly true that an increasing part of the collection of information that many libraries acquire, organize, provide access to, and preserve will be in electronic form, and as this proportion increases it will have wide-ranging effects on all aspects of library planning, operation, policy, and funding. How quickly this transition will move and how soon a critical mass of information necessary to serve various classes of library users will actually be available in electronic form are subject to considerable debate. I think that it will take longer than many people believe. There are a number of obvious situations in which much of the primary data in various areas is available and heavily used in electronic form, including space sciences, remote sensing, molecular biology, law, and parts of the financial industries. In some cases these data can only be meaningfully stored and used in electronic forms. But even in many of these areas much of the journal literature (as opposed to primary information) is still available only in print. Already there are massive data and information archives available on the network, containing everything from planetary exploration data to microcomputer software. While these are often referred to (rather grandly) as digital libraries, they are really not, in many cases, part of any actual library's collection and are not really managed in the way in which a library would manage them. In a number of cases they are actually the volunteer efforts of a few interested and energetic individuals, and there is no institutional commitment to maintain the resources.

It is important to recognize how little of the existing print literature base is currently available through libraries in electronic form, even to the limited and well-specified user communities that are typically served by the still relatively well funded academic research libraries. Libraries have been investing heavily in information technology over the past two decades, to be sure, but most of this investment has been spent on modernizing existing systems rather than acquiring electronic content. The most visible results of this program of investment are the on-line catalogs of various libraries that can be freely accessed across the Internet today (although they often contain a mix of databases, such as the catalog of the library's holdings, which are available to anyone, and other licensed information resources, such as databases of journal citations or full text, which are limited to the library's direct user community). These information retrieval systems allow a user to find out what print-based information a library holds; while they are very heavily used, their end result is normally citations to printed works rather than electronic primary information. Even a system such as the University of California, which operates a sophisticated system-wide information retrieval system and had been aggressively investing in electronic information resources for some years, only offers access to a few hundred journals and virtually no books in electronic format. It holds active journal subscriptions to over 100,000 journals and its on-line catalog contains entries for over 11 million books. (In many cases the journal articles are only available to the UC community a month or more after print publication; in most cases the electronic form of the article does not include graphs, equations, or illustrations.) The conversion of the major research libraries to electronic form is years away and is less a technological problem than a business and legal issue.

Interestingly, it seems likely that conversion of critical masses of electronic information may occur first for specific, often modest communities that are willing to pay substantially for access to the information resources and that are willing to pay personally (or at least organizationally), rather than through the support of an intermediary agency like a library. In these situations it also seems fairly common for individual users to be willing to pay for access to information transactionally (i.e., by the article retrieved or viewed or by the connect hour) rather than under the flat-fee license model favored by libraries. In many cases primary rights-holders feel much more comfortable with an arrangement that gives them income for each use of the material, rather than trying to decide "what it's worth" in advance as they would have to do in a flat-rate contract agreement. Indeed, this conversion of information resources to electronic form is already well advanced in the legal community. This area has moved quickly because it seems that the user community is not price sensitive and because of some unusual situations with regard to concentrated ownership of much of the key material by corporations that are both rights-holders and providers of access services directly. And, as has already been discussed, it may prove impractical or even impossible, for a variety of reasons, for libraries to provide much patron access to these electronic information resources once they are established. Certainly this has been the case in areas such as law (with the exception of law school libraries, which benefit from special arrangements intended to familiarize future graduates with the electronic resources).

It is important to recognize that these commercial information providers are not libraries, though they may offer access to immense databases that represent key resources in a given discipline. They acquire and mount information to make a profit and remove it if it does not generate sufficient revenue. They will not preserve little-used information just because it is an important part of the historical or cultural record. And many of these providers have followed a marketing strategy that emphasizes sales to a limited market at very high prices rather than a much larger volume market at much lower prices per unit of information.

In considering the conversion on the existing print base to electronic form, one basic issue that must be considered is the negotiation for license with the rights-holders. In most disciplines there are many publishers who contribute to the print base. The notion of a major research library negotiating and managing contracts with thousands of publishers is clearly absurd, as is the notion of negotiating a contract for a $100/year journal. (It will cost much more than that to negotiate the contract, in the absence of a "standard" contract that all parties could just accept unmodified. Such standard contracts do not exist today, and there has been little success in developing one, despite the efforts of such organizations as the Coalition for Networked Information.) A number of companies have begun to act as rights aggregators for libraries: University Microfilms Inc. and Information Access Company, for example, will license libraries' full-text databases containing material they have, in turn, licensed from the primary publishers. But coverage of these full-text databases is limited, and the material is not sufficiently timely to substitute for print subscriptions (or electronic access directly from the publisher). Until libraries can acquire electronic information under some framework that matches the simplicity and uniformity that characterizes the acquisition of most printed material today, the growth of electronic library collections will be slow.

The limited amount of library collections available in electronic form is already beginning to have effects that cause concern. Network-based access to electronic information is unquestionably convenient, and users tend to use electronic information resources that are available without consideration that they may be incomplete or may not represent the highest quality or most current work on a given subject. To some users, electronic library collections are already being viewed as defining the literature in a given area; these users have been too quick to embrace the transition to electronic formats.

In the next decade we will be dealing with a mixed transitional environment for libraries, where some parts of the collection exist in electronic formats but a large part continues to be available only on paper. This will greatly increase the complexity of managing the provisions for library services, defining priorities, and developing policies and strategies to position libraries to function effectively in their role as information providers to the public both in today's geographically bound, print-based world and in the future environment of electronic information on the NII.

CONCLUSION: THE BROADER CONTEXT OF
INFORMATION PUBLISHING ON THE NII

It is important to recognize that libraries are part of an enormously complex system of information providers and consumers that also includes publishers, government at all levels, scholars and researchers, the general public, scholarly societies, individual authors, various rights-brokers, and the business community. Even the much more constrained and relatively homogeneous scholarly communications and publishing system is vastly complex. Libraries play an integral part, but only a part. There may be an expanding division between the world of mass market information, much of which is bought and sold as a commodity in a commercial environment, and scholarly information, which is primarily produced and consumed by nonprofit institutions that place great value on the free flow of information (although there are currently many for-profit organizations involved in the scholarly publishing system).

There are vast changes taking place throughout this entire system as a result of the possibilities created by networks and information technology. Organizations are reassessing their roles: scholarly societies that once viewed themselves very much like commercial publishers are now rethinking their relationships with their disciplinary knowledge base, their authors, their readers, and the libraries that have traditionally purchased much of their output. Individual scholars are exploring the possibilities of various types of network-based publishing outside the framework established by traditional publishers, and in some cases also outside the customary social constructs such as peer review. Indeed, processes such as publication and citation that were relatively well understood in the print world and that are central to numerous areas in our society are still very poorly defined or understood in the network context.

Libraries are still basically oriented toward artifact-based collections and toward print. They are struggling to respond to the new variations on publishing and information distribution that are evolving in the networked environment and to determine what their roles should be with regard to these developments. Considering that libraries have played a relatively minor role in anything involved in television or radio in this regard, I would suggest caution in predicting their roles in a system of information creation, distribution, and use that is being transfigured by the introduction of networks and information technology based solely on the effects of these technologies on existing library operations and services rather than on the system as a whole. I do not believe that libraries are going to disappear, but they may change substantially in their roles, missions, and funding in the coming years. Certainly there is going to be a growing demand for properly trained, innovative, technologically literate librarians in many new roles in the networked information environment.

It is clear that the visions of information access through the NII are attractive and represent worthy societal goals. In many regards they build on the tradition of public libraries and their essential role in our society as it has developed during the 20th century. But without outright subsidy of access to electronic information or major changes in the framework that determines the terms and conditions under which the public has access to such information, these visions of the NII's potential may be difficult to achieve, and it may be unrealistic to assume that libraries have the resources to step up to these challenges. It seems clear that libraries have a number of potential—perhaps central—roles to play in implementing the public policy objectives that have been articulated for the NII. But they may not be the route to achieve all of these objectives, and in any event their roles and efforts will need to be complemented by investments from other sources, such as the federal government, in the development of public access information resources. We need to be very clear about who the user communities are, who is to be subsidized, what is being subsidized, and how that subsidy will be financed.

Discussion

ROBERT KAHN: Bob Lucky, if your model of the Internet being free is really going to hold for the rest of the country, how are we going to get it into the 70 million homes in America at $37,000 plus $7,000 each?

ROBERT LUCKY: I don't know. That is where the cost is going to be. The model is really based on business use, where we pay an average of $5 per person per month for everything that we can associate with the Internet, while the telephone might be closer to $100 when you throw in the long-distance charges and so forth. But in trying to get them into homes, particularly if there is a separate line involved, you face this enormous cost and bottleneck of that access, and so I simply do not know.

PAUL MOCKAPETRIS: I was struck by your comments about the difference between the phone company model and the Internet model. You said that the local loop is what is expensive and the long distance is not, in the phone company case. In the Internet case, you said you pay for access but then the Internet is free, so it seems like those two are the same model in the sense that the local loop is the thing that is dominating the cost, and I was just curious whether I was wrong or not.

LUCKY: I am not sure I understand, but it is sort of the same form as Bob Kahn's question. The costs are in that access. In business you have a model that is able to share that cost among many people because you have one T-1 line into a big building and then you have the local area network where most of the costs are. But most of us associate those costs in the local area with our computing environment and not the communication. I spent millions of dollars keeping up all the local area networks and all the people that run them and so forth, but I say, I have to do that for computers anyway and the communication in the Internet just comes out for free from that.

A totally different model applies for the home because it is one on one and I don't know how you would do that unless you overlay it on your voice. I mean, if you share that line and you have a flat rate, then again you could conceivably have Internet access that is rather cheap.

MOCKAPETRIS: I have one other question. A lot of people have different metrics about what part of the Internet is commercial and what part is not. What is the right way to count? We have numbers about how many hosts are registered that are not entirely sound in some ways—sort of like sampling license plates and seeing how many are commercial and how many are government. What would be the appropriate way to think about that? Packets or packet miles?

LUCKY: I do not know; we do not know. I do not think you can count packets and distinguish which is which right now, so again I am sorry not to be able to answer these questions. In fact, no one knows how many Internet users there are really. People say 20 million, but all we know for sure is the number of registered hosts. I have no idea, even in our company; I said 3,300 people. We have 3,300 addresses, but we think maybe only 1,200 are active. We really do not have good statistics on a lot of the stuff.

GEORGE GILDER: How many hosts?

LUCKY: About 2 million hosts.

GILDER: The concept of universal service confuses me. It took 50 years and about a trillion bucks to get universal service, something like that. To proceed forward with some requirement for universal service would stifle all progress. You cannot instantly create universal service. Universal service is something that can happen when you get sufficient volume so that incremental costs are very low. You have to start as an elite service almost necessarily. I wonder, when you speak of universal service, just what kind of requirement is meant or how that can be defined.

CHARLES FIRESTONE: I think that what you are going to see is a dynamic concept of universal service. You are right—something that is new and innovative but eventually becomes necessary or heavily penetrated might become universal service in a future year.

One of the big issues that we are going to be facing is to figure out what should be included in the bundle of universal service (just like we need to figure out what should be included in the bundle of privacy rights), and this, I think, right now is just a connection, a dial-tone connection. One of the issues, for example, is touch-tone telephone service. Is that something that in some states is considered universal service? People have to have access to touch-tone or they have to have a touch-tone connection into their homes, because otherwise they cannot avail themselves of the services I think you have in mind.

But eventually there may be some governmental service—I am thinking down the line—some access to local information, your community. I know, for example, that Santa Monica has had the PEN [Public Electronic Network] system. Now, universal service may only mean having free access to the Santa Monica City Council or their local network—an ability to access without having to pay extra for it. But it is something that has to be dynamic and I am sure will be.

The other question is the application to libraries. What is the library equivalent in the electronic era? How do you connect into information, and what information should be a public resource as opposed to something that is available strictly on a pay-by-the-drink or per-bit basis. That is something that I think we as a society we are going to have to come to grips with.

MICHAEL ROBERTS: There is a lot of clamor these days for improving the security of the network, especially the Internet, so I think Colin Crook has got it right. Especially in a network that has to have universal access, you assume the network is at some level insecure and you secure the applications. The question is, for public-sector areas that are critical, such as health care, libraries, and also intellectual property, what sort of process should we be thinking about from a policy standpoint in securing those applications? Is it possible, for instance, to have the sort of certification of applications that the financial people can do very privately and behind closed doors applied to private-sector applications? Is it possible to apply product liability sorts of considerations to that area even in the public sector?

EDWARD SHORTLIFFE: I have some thoughts about it. One of the intriguing problems about privacy and confidentiality of health care data is that the issue is not necessarily the security of an application per se but what somebody who has access to the data does with them. In other words, there is a potential for abuse by people who are authorized to access patient data. That is why legal remedies with criminal penalties are required when someone misuses privileged medical information to which they have access. There are no national standards in this area at present. As a result, one of the big emphases I have heard in the medical area is that we need to begin to introduce some uniform penalties, probably with preemptive federal legislation about the misuse of data to which people do have valid rights of access.

As for the more generic issue of trying to prevent people from breaking into data sets, instituting appropriate certification of software has a valid role. What you are talking about is simply having databases out on the network with lots of different applications that could access them. Of course, you run into the question of what should be the nature of the varied access methods that you could use to get to those data sets, independent of specific applications that may have been written to access them.

MARY JO DEERING: I have a question for Ted Shortliffe, whose clear vision I have always appreciated, but also for people in the audience who are on both the engineering and content side. It picks up on something that Charlie Firestone said this morning about the process of disintermediation that is going on in society with regard to information. The same thing is going on in health, actually, and it is a process that is parallel to the deinstitutionalization of health care. Health care reform is really going to continue pressure in those directions, with an emphasis away from hospitals and high-end acute care providers, toward primary care and preventive medicine and home care, for that matter.

My request specifically to Ted right now concerns that wonderful sketch [Figure 3], where you actually began to paint what the health information infrastructure would look like. It did not really include any of the linkages that would reflect that type of new reality in health care, that would provide the linkages that would be necessary among nonhospital institutions, nonspecialty providers, and perhaps the consumer. I think we would all like to see what that would look like.

SHORTLIFFE: You are absolutely right. As you know, we had a meeting on the subject of the NII and health care last week that the National Research Council sponsored ["National Information Infrastructure for Health Care," October 5-6, 1993] at which I think the message was driven home loud and clear. I personally am in primary care, and therefore I am sensitive to it as well. Good point.

KAHN: I have a two-part question, one part for Cliff Lynch and one part for Ted Shortliffe. I know, Cliff, that you are very well aware of the importance not only of accessing information but also of delineating what you can do with it—when you get it, not only whether you can copy it or distribute it or make derivative works, but all the things that are typically covered under copyright. In your talk you really did not get into that, and I am wondering what your own views are as to how that particular aspect of library development is going to proceed. How are we going to know what we can do with this stuff?

CLIFFORD LYNCH: I think that this issue of what you can do with information is already a major problem. As soon as you move into a mode where you are licensing information rather than purchasing it, and licensing information from different sources with different contractual provisions—and, just to add to the fun, do not necessarily have a standard sort of taxonomy of things you can do—you are into a situation where it is very difficult to know what you can do beyond reading the information once and then purging it from memory and never printing it.

I think that this is a particularly troublesome trend, as we start thinking about much more intelligent ways of handling information. As people start building personal databases and using intelligent agents, "knowbots," things of that nature, that correlate information from multiple sources and refine it, this sort of trend may be a major barrier to making intelligent use of information. It is something I am really concerned about. Another dimension of this is multimedia. I keep hearing about multimedia, but at least on bad days, I think the only ones who will be able to afford it are groups like major motion picture studios because only they can afford enough lawyer time to clear the rights.

ROBERT PEARLMAN: This issue is very, very big in the education sector, because one of the things that kids are doing these days is making multimedia reports, combinations of video and text and graphics, and they just grab images and text from everywhere, and it is a tremendous ethical problem.

But what has happened is that an industry is developing now, and CD-ROMs are out that basically provide images that you can use. In other words, an industry is growing up that says, "Sure, *National Geographic* will sell it to you at a big price, but we will give you similar images that you can buy and use for practically nothing," so there may be this kind of development in other sectors as well as competing sources of information.

KAHN: We have heard from Bob Lucky that the Internet, to certain kinds of businesses, essentially looks almost free and from Bob Pearlman that to the educational community it is virtually uneconomic at the moment. The other part of my question for Ted Shortliffe is where the technology fits in the case of the medical community. I think we heard from Ted that there is a strong motivation on the part of physicians in the hospitals to gain access to this technology. The medical profession itself, and I believe you, could provide a good, strong justification for that. In fact, I see it coming. The question I have is whether there is any economic justification for the end patient, namely, the user of the health care system, in having some coupling to the Internet and, if so, at what level of capability?

SHORTLIFFE: This relates to Mary Jo Deering's question. First, when you look at what the health care field is spending on computing right now, the amount of money that it would take for any given institution to hook up to the Internet is minuscule compared to what it is spending overall in information technology, so I don't see cost as a barrier per se. It is more the perceived benefit and how you actually would make use of the national network, given the lack of standards for actual connectivity and data sharing.

The same kind of argument could be made as we begin to see the evolution of new health care delivery plans. It may well become economically beneficial for health care providers to pay for linkages into the homes in order to provide patients with access to information that would prevent them from becoming more expensive users of the health care system. A lot of visits to doctors are unnecessary. If people only had more and easy access to the kind of information they might need, you can imagine how this might have an impact on overuse of certain kinds of facilities.

The problem with that approach is that you are hypothesizing not only an availability but also an education of the end-user patients and health care users which right now is unlikely. The biggest users of health care are the people least likely to have the facilities in their homes at present and the ones least likely to have the education that would allow them to make optimal use of such technology. So we are talking about a major social issue that would facilitate allowing patients in their homes, and people who are nonpatients who simply need access to health information, to make good use of the kind of information that might be made available.

KAHN: Is that a practical suggestion you just put out, that the doctors or hospitals literally pick up that mantle somehow?

SHORTLIFFE: It is not going to be the doctors. If you think the doctors are in charge of the health care system, you are a few years behind. It is going to be health plans as they begin to look at how they can compete for patients in large areas, especially in the big metropolitan areas. How this will play out in more rural areas is another matter that I think is a great worry to people, because the emphasis tends to be on the competitive marketplaces around the big cities. However, the health plans will pay for these technologies if they find it is to their competitive advantage to do so.

LINDA ROBERTS: I am trying to look for the common thread in all that we have heard this morning. It strikes me that in most cases what you have been talking about is doing what we do better. But I think there really is an opportunity to do better things in every one of the sectors that have been talked about this morning. I am thinking about the library as one example. We really have public libraries as a compromise that, if I understand it, was reached between the publishers and the communities that really wanted everybody to have their own library.

There were people who had libraries and who did not need public libraries, and they were the exception rather than the rule. What is so interesting about what could happen in the future is that there really could be a much more decentralized system of libraries. Everybody could have their own in their own home, and what is even more fascinating about some of the things that are happening in education when you talk to teachers is that a lot of the material that could be in these libraries does not necessarily have to be controlled and produced by the publishers out there.

So it seems to me that one of the things we really have to think about for the future is what we haven't been doing that we ought to be doing and that might create new opportunities for business and industry, but might also create new opportunities for learning more broadly.

VINTON CERF: With respect to something Bob Pearlman said, that somehow it was not economic for the schools to be a part of the network environment, I am a little puzzled because I assume that we spend, as a country, a great deal of money on education, so it is not like there are zero dollars out there. A lot of dollars are spent for education, so it is obvious a trade-off is being made about the utility of being a part of a network environment and spending money on other things. Maybe you can help us understand a little bit about how the educational dollar is spent. Is it mostly for personnel? Do we find networking not so useful until all of the teachers are trained in its use, as opposed to throwing equipment into the schools and expecting somebody to do something useful with it with no adequate software, and so on?

I want to suggest that we are like parents of teenagers when the kids are out late and we do not know where they are and we're worried. There are a thousand different possible things that could have happened to them even though it is probably the case that most of those things are mutually exclusive, and yet we still worry about all thousand of them.

We worry about how this technology will finally reach critical mass, will finally get to the point where there are enough people who have access to it for good, sound economic reasons that we can start doing some of the things that we have heard about this morning. I would like to suggest that we actually do not know yet which of the things will trigger the regular availability of all of the technology. The computer scientists got it first because they had to have the stuff to write programs. Then they got to do all these other neat things with the networks. Bob, what in your mind is the triggering event for making this economically interesting for the schools?

PEARLMAN: This, of course, is a very complicated question. If you look clearly at technology in U.S. schools, it has not resulted in many economies. For instance, in the mid-1980s there were some companies that were trying to market what were called integrated learning systems in the schools, on the basis that they would actually save in the use of teachers. In general they did not really save in the use of teachers, which is where the real economies are.

In schooling, about 80 percent of school budgets have to do with personnel, so the only way to reasonably save money is by saving on personnel. Some are saving on, say, custodial or food service personnel, but in the main you have to save on teacher personnel, and the only way you really get at that is by totally reorganizing schooling. I was associated with one of the new American school design awardee groups in Cambridge for the last year and a half. We came up with a design that we think in the long run is going to save money. It requires quite an up-front investment in communications technology.

For our kids to be able to work much more on their own with teacher advisors managing their affairs and with mentors, we had to establish up front an infrastructure in our design of a local area network in the school connected to the Internet with those up-front costs and with connections in the community, which meant that the community had to be wired properly. All of these up-front costs had to be borne by somebody.

With that kind of structure in place, we felt that all sorts of economies could occur in personnel in schools, but we had to test the proposition. The problem is that from the point of view of a school that exists right now, we are only talking about additional cost add-on. I don't mean that people really cannot afford to get onto a network; it is just hard for them to justify it. There are not real economies nationally. You get on the Internet, and what do you see? You are really just bringing in more programs—more kinds of things to look at, more curriculum possibly. But you are not really enabling a saving at that site.

What we really need is some real, serious innovation in the organization of schooling. That may take a lot of forms, and the charter school development may help. I am not a partisan of Whittle's efforts, but I am not happy that he cannot make it in the new design world. In fact, I would like to see some private player actually try to do it, because the question is how to put together the package, how to develop a school that is more economical when it is tied to a system—meaning many, many schools, whether in one location or around the country, that actually work together and produce some savings because they are sharing curriculum, curriculum costs, development costs, and teachers through the media. That is what has to happen.

Although we have things called school districts, or states that pretend to be school districts, they don't really bring about efficiencies in the way that a corporation like Citibank is trying to accomplish, looking at how to make a network that really makes all that blend together. So my answer is only that we are going to have to have lots of experiments and new designs for schooling over the next several years in order to actually exploit information technology properly.

CAROL HENDERSON: I want to pick up on an earlier question by Bob Kahn and tie together a couple of strains that we heard this morning in connecting the health care sector and the library sector. We often hear about patient records and telemedicine, but I do not know that it is widely realized how often people go from the doctor's office to the public library to try to get information about what it is they have just heard, what their child has been diagnosed with, or what they are looking at in terms of caring for their aged parents down the line, and so on.

When you look at the kinds of questions they ask, they really want to mine the medical field's information, but they want to do it, in a sense, outside the medical field because they want to know whether there are alternative kinds of treatment and what the literature says about this drug that they are supposed to be taking if they are also taking something else.

The usefulness of a neutral source of information about information is something that I think is very valuable, and perhaps we should not give up too easily. Access for patients, or people before they become patients, to preventive health care information and to information about their conditions is something that libraries can be a big help with.

I think perhaps we also passed over the idea of libraries as community institutions and as information providers. It is often the libraries that have mounted databases not just about their collections but also about community information and referral sources. Libraries are often the central source of information on where in a community you go for various government and social services across agencies as well as an institution that, for instance, reaches out to newer immigrant groups and helps get them into the mainstream.

Part 2

Regulation and the Emerging
Telecommunications Infrastructure

Introduction to Part 2

Roger G. Noll

National policy regarding the information infrastructure is often debated as if it were a novel idea; however, the current debate is in many ways old wine in new bottles. The backbone of the information sector of the economy—the telecommunications network—has been regulated by both national and state governments since the Bell system first perfected reasonably high quality long-distance interconnection in the first decade of the 20th century. From the beginning, at the center of the policy debate has been the importance of nationwide interexchange traffic to the welfare of consumers and the long-term economic growth of the nation. Thus, any attempt to implement a dramatic new initiative regarding the national telecommunications infrastructure must deal explicitly with the presence of an elaborate regulatory superstructure that, despite two decades of liberalization, still extensively controls the decisions of the most important players in the telecommunications industry regarding prices, investment, technology, and system integration. And, in the mid-1990s, many of the policy proposals for encouraging enhancements to the telecommunications infrastructure amount to recommended changes in regulatory rules and processes.

The purpose of the papers in this section is to examine the role of regulatory policy in shaping the evolution of the nation's information infrastructure, with special attention being given to how regulation affects the rate and pattern of technological change. In both public discourse and the scholarly literature in law, economics, and political science, the debate about the merits of regulatory policy deals with two quite separate and distinct issues. The first is philosophical and deals with the legitimate uses of the coercive power of government. The second is more prosaically practical and deals with the actual effect of regulation on the performance of industry. Although this section deals primarily with the latter, we must first briefly deal with the former in order to clarify the range of issues of concern.

REGULATION AND POLITICAL LEGITIMACY

The first focus of debate about the merits of regulation concerns the enduring philosophical question about the legitimate boundaries to the use of the coercive powers of government. Regulation makes rules about how people can spend their incomes, use their wealth, and engage in transactions, and so infringes upon private property rights. From this fact has emerged a debate concerning whether regulatory policy goes too far, or not far enough, in making a trade-off between individual liberty and collective welfare. Thus, this form of the debate about regulatory policy deals with the principles that should be adopted through a nation's constitutional and legal system to define the proper role of the state in economic affairs.

An important feature of this debate is that the instrumental value of regulation—its effects on the performance of regulated industries—plays a minor role. Causing the trains to run on time is not regarded as a compelling defense of fascism, for example, nor is the production efficiency of antebellum southern agriculture regarded as a serious argument against the Emancipation Proclamation.

More importantly, this form of the debate about regulation is most likely incapable of resolution regarding such prosaic issues as how, or even whether, government should attempt to control the development of infrastructural industries. Whereas the philosophical debate about the proper scope of state interference with private property rights does provide a compelling case against the more outrageous forms of authoritarian government, it does not contribute much to the debate in advanced western democracies about the appropriate methods for making and implementing economic policy. Among advanced western countries, for example, although no nation leaves the sector totally unregulated or has nationalized every aspect of it, national telecommunications policy does vary from near laissez-faire (New Zealand, United Kingdom) to an extensive nationalized enterprise (France, Italy). One cannot convincingly argue that any of these nations have organized telecommunications policy in a morally indefensible way. Hence, because the philosophical debate about the proper scope of government is, and is likely to remain, unresolved among the range of policies that are likely to be considered in a modern western democracy, this aspect of the debate about regulation is largely irrelevant and forms no part of the analysis in the remainder of this section.

INSTRUMENTAL ANALYSIS OF REGULATION

The papers that follow examine the second focus of debate over regulatory policy, which is concerned with the instrumental value of regulation.[1] At the most general level, the instrumental issue is how the presence of government supervision of an industry, regardless of the details, affects its performance. The existence of regulation redirects the time of a firm's management from ordinary business activities, such as production supervision, technological innovation, and customer relations, to dealing with and strategizing about political affairs and the regulatory process. An inherent feature of regulation is that, no matter how enlightened, it creates costs and affects the rate and pattern of technological change in an industry.

Some specific instrumental issues are how the performance of the regulated industry depends on the details of regulatory policy, such as the decisions about who will be regulated, what powers will be given to the regulators, how the regulatory authority will be organized and what procedures it will be required to follow, which level of government will be responsible for each element of regulatory policy, and what role will be assigned to the courts in overseeing regulatory policy. In the case of telecommunications, examples of specific issues that have been especially prominent in recent years are the debates about the jurisdictional separation of regulation of prices and entry between the Federal Communications Commission (FCC) and state public utilities regulators, the continuing judicial intervention in structuring the market for telecommunications services and equipment arising from antitrust litigation, and methods for regulating prices by a monopolist (e.g., rate of return, price caps, residual pricing, and so on).

The instrumental issues about regulation can be phrased in either a negative or a positive way. The negative version inquires about the extent to which regulatory policies and institutions distort the evolution of the telecommunications infrastructure, to the detriment of the welfare of society. The positive version seeks to identify ways that regulatory policy should be changed to improve the performance of a regulated industry. Of course, the way in which the question is put is primarily a rhetorical device, revealing more about the conclusion of the person addressing the issue than the actual nature of the debate. Regardless of how the issue is phrased, the core of the relevant policy debate is the following question: What will be the effect of different approaches to regulatory policy on the performance of regulated industry?

SOME GENERAL CAVEATS

By way of introduction to the papers that follow, the current debate about the national information infrastructure should be considered in the context of over a century of U.S. experience with regulatory policy in a variety of infrastructural industries. Economic regulation in the United States takes a peculiar form that is not found anywhere else in the world. It evolved as it did because of certain unique features of the U.S. political and legal system: the federalist system of government, the separation of powers among the three branches of government, the unusually strong role of the courts in the American system, and the specificity of certain rights in the U.S. Constitution, such as the guarantee of the sanctity of contracts, the protection against expropriation of property without compensation and due process of law, and the guarantees of free speech and freedom of the press. These special features of the U.S. system thwarted numerous attempts to establish economic regulation during the 19th century and continue to affect how regulation is implemented today.

The first important consequence of the U.S. system of government is that it makes policy making very difficult. A significant change in policy requires a statutory mandate, and statutes are very difficult to enact. Both houses of Congress and, in the absence of a two-thirds majority in either, the President, elected on the basis of different principles of representation, must agree before a statute can be passed. And, if someone then protests the new statute, the Supreme Court can declare it invalid—a power that was controversial at the time it was asserted by the court early in the 19th century, and that is shared by almost no other high court anywhere in the world.

The second important consequence of the U.S. Constitution is the elaboration of economic protection accorded citizens in the Constitution, as enforced and interpreted by an especially powerful court. Numerous statutes, mostly enacted by states, have been overturned by the Supreme Court because they were judged to interfere with one or another constitutional provision. Judicial skepticism of regulatory interference with the private economy persisted into the early years of the New Deal, when, under the threat of increasing the size of the Supreme Court to accommodate a more interventionist political philosophy, the "switch in time to save nine" finally permitted extensive interventionist policies. But even the post-New Deal courts placed important constraints on economic regulation. These constraints have been derived from both constitutional principles of economic rights and the details of statutes that elaborate how the constitutional role of the courts and protection of individual rights should be embodied in the details of regulatory policy making.

Constitutional and statutory requirements impose substantial burdens on regulators to prove that their decisions are justified. Regulators bear a legal burden to show that their decisions are based on a constitutionally valid statutory mandate, and an evidentiary burden to show that their decisions are rational means to pursue a statutory objective and take into account all of the relevant facts and arguments presented by those who might be affected by the decision. Again, the structure of the Constitution has the effect of making policy change slow and difficult, and subject to veto by the courts. Moreover, these requirements advantage large business interests with a major financial stake in regulatory outcomes, because they are more likely to provide detailed evidence in support of their interests than are consumers and other less well-organized user groups, and more likely to appeal adverse regulatory decisions to the courts.[2]

The history of telecommunications regulation illustrates these points. The origins of telecommunications regulation can be traced to the middle of the 19th century and the battle to establish railroad regulation.[3] Many states attempted to regulate railroads for the purpose of eliminating the "long-haul, short-haul" price differential in railroad prices. Typically, railroads competed for long-haul traffic but were monopolists at intermediate rural terminals along their networks. Railroads exploited these local monopolies by charging far higher prices for shipments to and from these rural locations than for more competitive transportation between large urban centers. Except for a brief period between 1877 and 1886, the Supreme Court persistently declared as unconstitutional these attempts by states to deal with short-haul monopolies.

In 1887, after the 1886 *Wabash* decision in which the Court declared all state regulation of any aspect of an interstate railroad to be an unconstitutional interference with interstate commerce, the federal government enacted the Interstate Commerce Act (ICA). This statute had two important elements. The ICA established federal regulation of railroads, and it gave states the authority to regulate railroad tariffs for shipments between terminals in the same state.

Two decades later, the scope of the ICA was expanded to include telecommunications.[4] The Mann-Elkins Act of 1910 enabled the Interstate Commerce Commission (ICC) to regulate the interstate portion of telecommunications; however, this business was then so inconsequential that the ICC ignored its new responsibility. The important effect of this extension of the statute was that it legalized state regulation of intrastate telecommunications. Then, in 1934, the Communications Act created a new institution to regulate the interstate portion of the industry, the Federal Communications Commission. But most of the telecommunications sections of the 1934 act were transferred wholesale from the ICA, despite many failed attempts within Congress to alter these provisions to match the differences in the technologies and market structures between railroads and telephones.[5]

In the ensuing decades, Congress has made numerous attempts to rewrite the Communications Act to reflect the realities of new technologies and changing market structures in the industry. Except for a handful of exceptions (laws dealing with satellites, cable television, and radio telephony), these attempts have failed. In 1994, Congress again failed to pass a reform bill (S. 1822, sponsored by Senator Ernest F. Hollings) that died in the Senate after passing overwhelmingly in the House. Consequently, the 1934 statute remains the basis for almost all telecommunications regulation today, including the scope and methods of regulation by the FCC and separation of authority between the FCC and the states. Thus, as the United States attempts to make policy about the information infrastructure of the 21st century, it does so in the context of a regulatory statute that was written in the 19th century to deal with local railroad monopolies.

The significance of the U.S. courts is clearly evident in telecommunications regulatory policy. Most of the important changes in telecommunications regulation since the passage of the Communications Act of 1934 are the result of court decisions, rather than new statutes or policies adopted by the FCC. Examples are the *Execunet*[6] decisions, which led to competition in ordinary long-distance telephone service, the restructuring of the industry in the settlement of *U.S. v. AT&T*,[7] and the *Louisiana*[8] decision, reversing 50 years of precedent to give more regulatory authority to the states over aspects of the local exchange that have a significant effect on interstate elements of the industry.

The preceding history is crucial to understanding why the academic literature is so skeptical about the value of economic regulation, both in general and with respect to telecommunications. As an instrument of public policy, regulation—at least in the U.S. system of government—has inherent properties that limit its effectiveness. Regulation is intrinsically slow, and changing regulatory policy is extremely difficult. The U.S. system of government exalts the status of individuals and private property like no other, and so imposing costly rules on anyone is difficult. In the United States, policies are developed and changed by an elaborate process that is designed to protect individuals against significant, targeted economic harm by government, and to allow change only when a broad consensus supports it. Thus, regulation is especially problematic as an institution for channeling the development of an industry in which technology is evolving rapidly and involves an ever-changing cast of companies and user groups.[9]

The fact that the "information superhighway" of the 21st century is being constructed under the aegis of a 19th-century statute for regulating railroads is not the result of stupidity, corruption, or inattentiveness on the part of government, but a highly significant example of a fundamental feature of U.S. governance. The United States operates under a system that is based on a deep skepticism of government intervention in economic affairs and that makes rational national planning of industrial development extremely difficult.

It is important to note that making industrial policy difficult and cumbersome, and hence less extensive, is not necessarily bad. To illustrate the point, consider the differences in the performance of the telecommunications sector in Japan and the United States[10] In Japan, judicial supervision and authority over economic policy are far less extensive, and legislative and executive functions are not formally separated. As a result, the legislature can more easily change policies by statute, and the implementing bureaucracy is less constrained about the scope of its decisions.

In 1986, the Japanese passed new laws that in many ways are very similar to the combination of court decisions and regulatory policies that took place in the United States in the previous two decades. By statute the Japanese created a structurally competitive telecommunications industry, but they retained their traditional system of ubiquitous government control over prices, investment, and entry into facilities-based telecommunications services. As a result, the number of facilities-based domestic telecommunications companies grew from 1 to 38 in 6 years, counting all satellite, wire, and over-the-air carriers.

Because the core economic decisions of the industry continued to be controlled by the government, the new structural competition did not lead to real economic competition. The new entrants were permitted to charge much lower prices for dedicated circuits, which are typically purchased by very large, multifacility businesses; however, in other aspects of the industry companies were prohibited from competing on the basis of either price or the quality of service. The objectives of this policy were in part to allow more companies to participate in the lucrative telecommunications services business, and in part, like U.S. policy objectives, to encourage more rapid development of a ubiquitous, integrated, high-performance telecommunications infrastructure.

By the criteria most Americans would use to evaluate the success of the Japanese policy change, the outcomes are not attractive. In 1992, a Japanese customer of telephone service paid about $560 for the installation of new telephone service, which is about 10 times the price charged for initial residential installations in the United States. Consequently, the number of telephones per capita in Japan is about 20 percent lower in Japan than in the United States. Thus, the more extensively regulated, less competitive system does a significantly worse job of achieving universal service.

After paying $560, a Japanese residential subscriber paid about $20 a month for basic access service in 1992, but this service does not include any calling. In addition, a Japanese consumer paid 8 cents for each 3-minute interval of a local telephone call, a charge that has since risen to 15 cents. In addition, the Japanese consumer faces significantly higher prices for long-distance calls, especially for calls at distances over 100 miles. As a result of this pricing system, the Japanese use their telephone system much less intensively than do Americans. Minutes of use per line per year in Japan are about one-sixth the amount of usage in the United States. The Japanese place many fewer telephone calls and have much lower average connect time than is the case in the United States.

Higher prices, lower penetration, and less usage in Japan obviously result in lower consumer welfare; however, the problems with Japanese telecommunications policy go beyond these effects. If an important social objective is to facilitate the introduction of new telecommunications services, the Japanese policy thwarts them. New services require more calls and connect time, and their ubiquitous availability requires universal service. The Japanese system accommodates neither, and so serves to retard the evolution of the Japanese information infrastructure in comparison to that in the United States.

Japan differs from the United States in many respects other than the legal and institutional structure of industrial policy, and so broad conclusions drawn from these facts need to be strongly qualified. Nevertheless, a slow-moving but limited regulatory system, coupled with greater reliance on competition, actually produces relatively good performance in comparison with the systems in other advanced, industrialized nations.

IMPLICATIONS FOR EVALUATING REGULATORY STRATEGIES

Of course, the preceding observations do not fully resolve the debate about regulation. Regardless of the problems associated with regulation, both citizens and government officials will seek to interfere with private economic decisions if they believe that these decisions do not adequately take into account important collective values. Although regulation has inherent features that limit its effectiveness, these effects are only one part of a larger concern. Inherently imperfect regulation can be better than nothing when dealing with a firmly entrenched monopoly or an economic activity that imposes significant social costs, such as creating pollution.

This introduction is intended to make a simple point that is often overlooked in the debate about regulatory policy: citing a meritorious policy objective is not sufficient to justify the conclusion that regulation is warranted. Moreover, even a strong case for a serious market failure leaves another important issue unaddressed: how to design the details of regulation to ameliorate to the maximum feasible extent the inherent infirmities of the regulatory process. Questions about how best to divide jurisdiction between federal and state authorities, or between regulators and the courts, and about what principles to adopt in controlling prices and service attributes, must be answered in part on the basis of their feasibility in the U.S. legal and political system. Because the process of policy making is important in the American system, the instrumental value of regulatory policies is often strongly influenced by the details of its implementation.

An important illustration of these points is the ongoing debate about cable television regulation. In 1992, Congress enacted a statute, over presidential veto, to reregulate cable television. A veto-proof two-thirds majority was obtained by constructing an elaborate statute, carefully defining what could and could not be regulated, how regulation was to be implemented, and who was to implement each aspect of regulation among the FCC, states, and local government. A year later, the FCC implemented the statute by issuing detailed pricing rules and a process for certifying local regulation.

On the basis of initial studies, the effects of this elaborate process appear to have been minimal. Some prices went up, others went down, but on average the changes were small. Most local governments have decided not to attempt to become certified regulators, regarding the cost as not worth the candle. The main observable effect has been a reshuffling of the channel numbers assigned to various programmers, a step necessary for most cable systems to minimize the effect of the new rules on their total revenues from subscribers.

At the heart of the new cable regulations was an important economic fact: cable television systems are enormously profitable, selling for several times their construction costs, because typically they have considerable monopoly power. But the question posed by the history of the construction and implementation of the 1992 statute is whether any cable regulatory system that is politically feasible and constitutionally valid can yield benefits that exceed its costs. Thus far, the preliminary results indicate that the answer is no, even though cable does enjoy revenues that substantially exceed the economic costs of service.

The problem with cable regulation is not the validity of the objective, but the thus-far unsolved problem of implementing a regulatory system that can make significant progress in achieving that objective. Of course, the cable lesson does not translate directly to telecommunications. Cable is a far less important industry than telecommunications, at least for now. Unlike telecommunications, cable is not an important factor determining the performance of many other rapidly growing, high-technology industries, and for consumers cable does not yet offer the highly valued, almost indispensable services that are accessible only through the telephone. Thus, the prospective benefits of telecommunications regulation are larger and would justify considerable inefficiency in an imperfect regulatory system.

Nevertheless, the cable debate makes clear the preference for competition rather than regulation that is expressed by numerous observers of the communications industry. The experience with cable adds weight to the argument that easing the entry of direct broadcast satellites, of off-air digital distribution systems, and even of second cable companies or telephone companies into the cable business will do a better job than regulation in holding down cable prices and expanding service offerings to consumers. Likewise, the cable experience raises important questions about whether the promotion of competition, perhaps accompanied by some direct, targeted subsidies for particular users, will do a better job than regulation of encouraging the development of the information superhighway of the 21st century.

Congress, the FCC, the Justice Department's Antitrust Division, and the courts have before them numerous proposals to restructure telecommunications regulation. The failed Hollings bill combined with alternative proposals from Republican Senate leader Robert Dole will continue to be live legislative issues in the next session of Congress. Two competing visions of the future are embodied in these legislative proposals. One (Hollings) formalizes the importance of competition policy by placing a burden of proof on Bell operating companies for having their line-of-business restrictions removed, reestablishes the pre-*Louisiana* preemption authority of the FCC, and legislates a broad universal service requirement that, among other things, requires all telecommunications carriers to subsidize service to various educational and nonprofit institutions. The other (Dole) imagines a smaller role for federal intervention, eliminating the line-of-business restrictions without further requirements, and all but deregulating interstate service and interconnection among carriers.[11] Nevertheless, these two bills share two important policies. First, both proposals would establish a tax on all telecommunications services for the purpose of subsidizing local exchange service in smaller communities. Second, both would eliminate restrictions against entry into local exchange service, including prohibitions against entry by cable television and other utility companies. (In Japan, electric utilities have been an important source of entry into facilities-based telecommunications services.)

Meanwhile, significant restructuring proposals have been submitted to the Antitrust Division, the FCC, and the D.C. District Court that oversees the divestiture decree. These proposals involve the movement of local exchange carriers into long distance and manufacturing, and of long-distance carriers into local access.

In the summer of 1994, the Antitrust Division allowed AT&T back into the local access business by allowing the merger of AT&T and McCaw Cellular, the nation's largest radio telephone company, after insisting that McCaw provide equal access to all long-distance carriers and that AT&T continue to supply cellular system equipment to McCaw's competitors on a nondiscriminatory basis. Later in the fall, implementing 1992 legislation that will eventually allow a competitive radio telephone industry, the FCC auctioned new spectrum for radio telephone services, which is expected to lead to the entry of several new access providers in all major metropolitan areas.

Facing some access competition already in the downtown areas of large cities, and much more potential competition from radio telephony and, conceivably, cable television, some local exchange carriers have proposed deals that would allow them to enter manufacturing and long distance, both of which would increase the chance that they could be more effective competitors in markets for information services. The most imaginative of these proposals comes from Ameritech, the regional Bell operating company in the Great Lakes region. Ameritech proposes complete unbundling of the elements of local service and relaxing all regulatory barriers to the entry of competitive access providers in return for eliminating its line-of-business restrictions. Unbundling, accompanied with realistic, cost-based prices, would enable competitors to enter the local access market by leasing and reselling elements of the Ameritech network in combination with their own facilities. For example, MCI might lease copper wire connections between households and the local central office, but at that point branch the connections to its own local switch rather than use Ameritech's switch, and might provide metropolitan interexchange service by building some of its own trunks between switches, but leasing some trunk capacity from Ameritech.

All of these proposals share a new vision of local access: because of encouraging prospects for radio telephony, independent local networks (most private, but a few public), cable television, other utility companies, and extension of the reach of long-distance carriers, local access ought not to be a protected monopoly, and may some day be reasonably competitive—and, perhaps, even unregulated. Disagreements arise about exactly when each type of company should be permitted into which markets and under what conditions, and about how to define, and to pay for, universal service. But the present reality is that more than half of investments in telecommunications networks are now being made for local systems other than the traditional monopoly local exchange network, and the political system, at least at the federal level, has pretty much reached accord that this diversity should be facilitated rather than retarded.

VARYING VIEWS ON REGULATION

The papers that follow provide a range of views about these developments, the prospects for competition in all elements of the industry, and the ways that federal and state regulators are responding to them. Robert Crandall summarizes the objectives and performance of telecommunications regulation. He points out that the objectives of regulation are more complex than simply protecting consumers against monopoly, and in particular that the objective of fairness—all citizens should have access to approximately the same range of services at roughly the same prices—conflicts with giving service providers proper incentives for operating efficiently and adopting warranted new technologies. Because of this conflict, regulation often is the primary barrier to competitive entry, made so because regulators fear that entry will upset the fairness of the system. Crandall explains that this sacrifice of efficiency for fairness is becoming increasingly costly as telecommunications technology progresses, and he advocates regulatory reforms that minimize the scope of regulation and, where regulation is necessary because of market power, that accord greater weight to economic efficiency.

Robert Harris picks up Crandall's themes about the growing importance of an efficient telecommunications system and the desirability of less extensive regulation. According to Harris, telecommunications must be seen as part of an overall strategy for economic growth, much as railroads were a primary engine of economic development in the 19th century. The rationale for government intervention is gradually shifting from protecting consumers against monopoly to assuring complete interconnection, including among competitive carriers. The latter concern arises because the network is more valuable to each user as the number of accessible people and businesses increases. Harris concludes with a survey of regulatory reform activities in several states that appear to be placing a greater emphasis on efficiency, and in some cases to be proactively procompetitive.

Dale Hatfield addresses the issue of competition between local telephone and cable television companies. Because of fundamental structural differences in the nature of these two local networks, Hatfield concludes that neither is likely to become an important competitor to the other in the next few years. Hatfield expresses concern that the movement by Bell operating companies into video services will be expensive and will be driven more by the distorting effects of regulation than by any efficiency or procompetitive advantage flowing from such entry.

Nina Cornell focuses on the precise nature of the bottleneck monopoly enjoyed by local access providers. Cornell's main theme is that as long as local exchange carriers have control over termination, they will have market power over the entire network. And, according to Cornell, historical experience teaches that control over termination has inevitably led to actions to extend the monopoly to other elements of the industry. Thus, for a procompetitive, free-entry regime to work, policy must succeed in making both originating and terminating access competitive. To do so will require unbundling of termination and efficient pricing of the unbundled elements. Moreover, the pricing regime will require more than price-cap regulation, since price caps do not prevent, and sometimes reward, strategic pricing to harm competitors.

Thomas Long expresses the common concerns of consumer advocates about the diversification of telephone companies and the adoption of efficient pricing systems for local access. The essence of Long's concern is that network upgrades for the few or that may not be efficient for anyone will lead to higher prices for all consumers. Reliance on access competition to prevent this outcome is, in Long's view, unrealistic, and regulators seem prone to believing that there is more competition than in fact exists. Long believes that allocating some of the fixed costs of the local network to services other than basic access is the best safeguard against consumers becoming involuntary investors in high-technology services that they do not want or need.

Bridger Mitchell's paper clarifies the economics of local telephone networks, and in particular the concepts of subsidy that are relevant for policymakers. Cross-subsidy occurs when, given current prices, other customers and the network supplier would be better off if a service were abandoned. Basically, this means that a service is not being subsidized if its price exceed the incremental cost of service. Mitchell than examines the cost structure of California local access companies to ascertain exactly who is and is not subsidized. Mitchell's work indicates that:

- Flat-rate residential service is probably economically efficient (the gains from measured rates are more than offset by the measuring and billing costs);

- Because local access costs are driven primarily by the distance of customers to the local switch, state-wide rate averaging subsidizes customers in sparsely populated areas and discourages the efficient adoption of radio telephony rather than wirelines as the means of providing access in rural areas; and

- The incremental cost of service is less than half of the average cost, indicating that in most areas wireline access is a natural monopoly and that most residential subscribers are not being subsidized.

Eli Noam provides a conceptual framework for understanding how regulatory issues evolve over the life of a communications network. Initially, because of economies of scale and the fact that each consumer values the network more highly as it acquires more customers (the "network externality"), the primary issue is promoting the growth of the system. This phenomenon applied to basic access for the early history of telephony, and applies today to new capabilities such as Internet and, perhaps, digital transmission. Once these scale and externality factors become less important, the primary policy issues shifts—to promoting competition and mandatory interconnection where warranted, but to protecting "cream-skimming" entry driven not by efficiency but by price distortions arising from the "fairness" objective. Finally, in the late stage of a system, costs are rising and further expansion of the system is not worthwhile, and policy focuses more on encouraging migration to another, usually more advanced system and making certain that those with investments in the old system do not succeed in artificially maintaining its dominance by erecting barriers to entry—including regulatory, legal, and political barriers.

Collectively, the papers in this section reflect the dilemma of regulatory policy as technology and market structure rapidly evolve in the telecommunications industry. The core economic uncertainty is how important the new technologies will be, whether that importance is limited to a minority of sophisticated customers or a large fraction of the businesses and residences in the nation, and the extent of natural monopoly that will be present in each major component of the network during the next decade or two as the use of these new technologies spreads. The political problem flowing from this economic uncertainty is the relative importance of the two major duties of regulatory policy: protecting consumers against a ubiquitous, impregnable monopoly and assuring universal access to prosaic telephone service, or facilitating the rapid adoption of advanced new technologies. For the most part, conflicts over the proper nature and scope of regulation, including whether extensive deregulation is in order, arise from continuing disagreement about the technological and economic future. But in setting a future course, policymakers need to take into account that a regulation-intensive process tends to equalize prices and access across groups, but in the process tends to slow change, inhibit the entry of new firms with new ideas, and advantage status quo players.

REFERENCES

Brock, Gerald W. 1981. *The Telecommunications Industry: The Dynamics of Market Structure.* Harvard University Press, Cambridge, Mass.

Capron, William. 1971. *Technological Change in Regulated Industries.* Brookings Institution, Washington, D.C.

Cass, Ronald A. 1989. "Review, Enforcement, and Power under the Communications Act of 1934," *A Legislative History of the Communications Act of 1934,* Max D. Paglin, ed. Oxford University Press, Oxford.

Joskow, Paul L., and Nancy L. Rose. 1989. "The Effects of Economic Regulation," pp. 1449-1506 in *Handbook of Industrial Organization,* Richard Schmalensee and Robert Willig, eds. North-Holland, Amsterdam.

Kanazawa, Mark T., and Roger G. Noll. 1994. "The Origins of State Railroad Regulation," pp. 13-54 in *The Regulated Economy,* Claudia Goldin and Gary D. Libecap, eds. University of Chicago Press, Chicago, Ill.

Noll, Roger G. 1989. "Economic Perspectives on the Politics of Regulation," pp. 1253-1287 in *Handbook of Industrial Organization,* Richard Schmalensee and Robert Willig, eds. North-Holland, Amsterdam.

Noll, Roger G., and Frances M. Rosenbluth. 1995. "Telecommunications Policy: Structure, Process, Outcomes," *Structure and Policy in Japan and the United States,* Peter Cowhey and Mathew D. McCubbins, eds. Cambridge University Press, New York, forthcoming.

Poole, Keith T., and Howard Rosenthal. 1994. "Congress and Railroad Regulation: 1874-1887," pp. 81-120 in *The Regulated Economy,* Claudia Goldin and Gary D. Libecap, eds. University of Chicago Press, Chicago, Ill.

Robinson, Glen O. 1989. "The Federal Communications Act: An Essay on Origins and Regulatory Purpose," *A Legislative History of the Communications Act of 1934,* Max D. Paglin, ed. Oxford University Press, Oxford.

Thierer, Alan D. 1994. "Senator Dole's Welcome Proposal for Telecommunication Freedom," *Heritage Foundation Backgrounder,* No. 233 (August 24).

NOTES

1. For an excellent summary of the research literature on the effects of regulation on economic performance, see Joskow and Rose (1989).
2. For a survey of the research that explores how politics and political institutions shape regulatory policy, see Noll (1989).
3. For more about the origins of railroad regulation, see Kanazawa and Noll (1994) and Poole and Rosenthal (1994).
4. For a good history of the early telecommunications industry and its regulation, see Brock (1981).

5. For details about the close statutory relationship between the Interstate Commerce Act and the Communications Act of 1934, see Cass (1989) and Robinson (1989).

6. *MCI v. FCC*, 561 F.2d 365 (D.C. Circuit 1977) and 580 F.2d 590 (D.C. Circuit 1978).

7. The preliminary ruling by Judge Harold Green rejecting summary judgment and indicating a likely victory for the government: *U.S. v. AT&T*, 552 F. Supp. 131 (District Court for D.C. 1982); the final judgment divesting the company (entered as a modification of the settlement of the earlier case): *U.S. v. Western Electric*, 569 F. Supp. 990 (District Court for D.C. 1983).

8. *Louisiana Public Service Commission v. FCC*, 476 US 355 (1986), held that the FCC could not assert jurisdiction over intrastate regulation simply because the latter affected the FCC's interstate policies if it was technically possible to divide jurisdictional authority. This decision explicitly reversed the previous judicial interpretations of FCC authority, which held that the FCC could assert jurisdiction if a service was not wholly intrastate in character, as most clearly articulated in *North Carolina Utility Commission v. FCC*, 537 F.2d 787 (4th Circuit 1976). The reach and durability of the *Louisiana* decision remain disputed and controversial, but the decision has certainly generated considerable inconsistency in the jurisdictional separations between state and federal authorities. For example, the FCC was allowed to assert jurisdiction over telephone instruments for the purpose of unbundling and deregulating them (see *North Carolina* and *Computer and Communications Industry Association v. FCC*, 693 F.2d 198 (D.C. Circuit 1982)), but when it attempted to do the same thing for inside wire it was allowed to require unbundling but not to require deregulation (*NARUC v. FCC*, 880 F.2d 422 (D.C. Circuit 1989)).

9. For an excellent compendium of how regulation altered the rate and direction of technological progress in infrastructural industries during the 1950s and 1960s, see Capron (1971).

10. For more details about the comparison between Japan and the United States, see Noll and Rosenbluth (1995).

11. For a descriptively accurate if somewhat partisan analysis of the two basic approaches, see Thierer (1994).

Government Regulation and Infrastructure Development

Robert W. Crandall

The 1970s and 1980s provided students of regulation with a wealth of empirical evidence on the effects of regulation and deregulation on prices, output, and service quality in many industries. This evidence has confirmed much of the prior research that concluded that regulation reduced economic welfare. Indeed, Winston's recent comprehensive survey of the field suggests that, if anything, economists underestimated the benefits of deregulation, in large part because they failed to predict the development of new services and technology after deregulation (Winston, 1993).

It has been 34 years since the Federal Communications Commission (FCC) began to liberalize the telecommunications sector by allowing private microwave service; 24 years since the FCC allowed entry into private-line long-distance service; 19 years since MCI entered switched long-distance service despite the FCC; and 11 years since AT&T agreed to a divestiture to settle a 1974 antitrust suit. Communications technology has improved at an explosive rate over this period, and there appears to be no slowing of this progress on the horizon. Despite all of these changes, telecommunications remains a highly regulated industry—regulation that is still justified by concerns over natural monopoly.

If, as Winston found, economists were unable to predict how the airlines would build their networks after deregulation with a fleet of jets whose technology has changed very little since the 1978 deregulation, economists should certainly be wary of pretending to know how the nation's communications infrastructure will develop with or without the heavy hand of government regulation and/or subsidy. While some students of telecommunications technology—many of whom are potentially heavy users of new network infrastructure—may have strong views about how they would like to see the nation's communications infrastructure develop, their vision may not be consistent with the developments that maximize economic welfare.

It is my view that regulation should not be seen as a form of government infrastructure planning, but rather as a source of restraint on actors who enjoy some modicum of market power that should decline as the potential for monopoly diminishes. In this paper I attempt to outline the major regulatory issues that have arisen in telecommunications as some regulators have moved to liberalize market entry while still maintaining control over a variety of rates and other operating parameters of incumbent carriers. In so doing I shall focus as much as possible on the implications of these issues for infrastructure evolution. I start, however, with a brief digression on the definition of "infrastructure."

WHAT IS INFRASTRUCTURE?

To many participants in the debate over industrial policy, infrastructure is generally thought of as those facilities owned by government or by private, regulated utilities. By this definition, common-carrier telephone systems, common-carrier energy distribution systems, sewers, water utilities, government buildings, and public schools are infrastructure, but private computer networks, private schools, private roads, or shopping centers are not.

This distinction between public capital and private capital in unregulated firms may be useful for some purposes, but it is surely misleading in telecommunications. Billions of dollars have been invested in private networks that are at least related to if not part of our communications infrastructure even though they are not generally available to other users. Presumably, these facilities reflect the decisions by private concerns that total reliance on common-carrier facilities is not in their best interests. Indeed, it may be that the role of common carriage is rather limited in an efficient, dynamic market with rapidly changing technology and equally rapidly changing user requirements. It is even possible that the only communications infrastructure may eventually be a set of interfaces among myriad private networks.

Unfortunately, there may be no way for the student of telecommunications to determine which combinations of private and common-carriage network facilities reflect an optimal config-uration. It is for this reason, among others, that public policy is moving toward more open access of unbundled common-carrier facilities, allowing a myriad of possible combinations of common-carrier and other facilities.

In the discussion that follows, I assume that the goal of public policy should be to allow a multitude of actors to invest in new technology and to develop the public/private network in a fashion consistent with their needs. This may result in a multitude of private networks that interconnect in some hierarchical fashion with or without the assistance of large common carriers, or it could lead to a number of competing large networks—owned by, say, AT&T, MCI, TCI, Time-Warner, Hughes, and the regional Bell operating companies.

WHERE IS THE NATURAL MONOPOLY?

The stated objectives of most telecommunications regulation is the protection of the public from the evils of monopoly power that might be exerted by telephone common carriers through raising rates above incremental costs and/or denying their actual or potential rivals access to their bottleneck facilities. In fact, the objectives of most regulators are much broader than this. Operating in a political arena, these regulators view regulation as an exercise in "fairness," protecting residences, small businesses, and rural Americans from having to pay the full (long-run incremental) cost of basic telephone service.[1] As a result, it is widely accepted that other services "subsidize" basic local service for these groups of ratepayers.

Some may and will debate this conventional wisdom about widespread cross-subsidization in telephone regulation. However, one recent empirical study confirmed the existence of such cross-subsidies (Palmer, 1992) using Faulhaber's test for subsidization (Faulhaber, 1975). Many others have reached the same conclusion through more informal empirical tests (Perl, 1985; Crandall, 1991).

Even if cross-subsidies do not exist in the sense that rates do not lie outside the range between stand-alone costs and incremental costs, it seems clear that state regulators in particular have historically been quite reluctant to allow prices to move toward Ramsey (quasi-optimal) levels (Baumol and Bradford, 1970). Specifically, usage rates—particularly for toll calls—have been kept too high and access rates too low for most customers in response to apparent political demands.

Ironically, it has always been thought that the most likely locus of natural monopoly in telecommunications is in providing access to dispersed small customers. But there cannot be any test of the existence of such power as long as regulators deliberately keep the price of access artificially low to satisfy political objectives while frustrating entry into other markets from which the "subsidies" are paid.

Recent changes in technology have cast doubt on the notion that local access markets are natural monopolies. The development of new access technologies that employ the electromagnetic spectrum are threatening to provide ubiquitous competition for the traditional wire-based systems. Cellular competition is increasing as providers seek out new spectrum and employ new digital technology. Personal communications networks are only in their infancy, but they too may add substantially to consumers' choices for gaining access to the network.

On the other hand, fiber optics may have such large-scale economies as to threaten to reestablish AT&T's long distance monopoly (Huber et al., 1992).[2] Now that there are three national long-distance networks and numerous other regional carriers, one might have expected interstate long distance services to be deregulated by the FCC. But fear of AT&T dominance and evidence that competition has not been responsible for lower rates (Taylor and Taylor, 1993) have made such deregulation difficult to complete. Fortunately, even if minimum efficient scale in fiber optics transmission is very large, transmission costs may be such a small share of total long-distance service costs that such economies alone could not reestablish AT&T's monopoly.

All of the current discussions about the locus of natural monopoly in telecommunications avoid a central fact: the potential monopoly power that exists today is as much a reflection of regulatory barriers to entry as to technology and market conditions. We simply do not know whether such power would persist in the absence of regulation because we have no market experiment to instruct and guide us. Moreover, given the incredible rate of technical change in this sector, we have no reason to believe that assertions based on today's array of telephone poles, paired wires, cables, and switches are very informative. Somehow, we need a market test of the effectiveness of competition in this sector.

THE REQUISITES OF PUBLIC POLICY

Given the rate of progress and the large number of actual and potential players in this sector, it is my view that regulatory policy should now stress:

- Removal of the worst of the remaining rate distortions for access and usage;
- Elimination of entry barriers into local, interexchange, and "information" services markets;
- Deregulation of all but the "core monopoly" services and substitution of price regulation for rate-of-return regulation for these services; and
- Regulatory forbearance from any attempt to guide network design or investment decisions.

Removal of Rate Distortions

There is no way that regulators can know the "true" costs of various telecommunications services given the pervasive joint and common costs in this sector and the rapid rate of technical change. Allocations of historical accounting costs are quite simply arbitrary exercises. Simulations of costs through forward-looking engineering models are equally subject to error and abuse by regulatory participants.

Nevertheless, it is fairly clear that toll usage rates are undoubtedly too high because regulated carrier access rates are kept above long-run incremental costs. Moreover, it is equally clear that the incremental cost of a business loop is generally little different from the cost of a residential loop in the same geographical area and that the cost of access increases with the distance from the switch. Thus, a reduction in business access rates, an increase in residential rates—particularly in smaller communities—and a reduction in carrier access rates are generally required to move rates closer to incremental costs.

As rates are moved toward incremental cost, competition in providing local access to smaller customers will increase. At some point, cellular, personal communication networks (PCNs), or other technologies might actually begin to make local access a contestable market. Once this occurs, the degree of regulation of local exchange carriers (LECs) can be reduced, and LECs can be freed to compete in vertically related markets.

Removal of Entry Barriers

State regulators have historically been reluctant to admit competitors into local markets for fear of undermining the cross-subsidies they have crafted. Not only will elimination of cross-subsidies increase the probability of competition, but increasing competition will reduce the possibility for cross-subsidies.

Entry is taking place. State regulators are finding it increasingly difficult to prevent new fiber optics networks from developing in large cities, and they are largely powerless to stop the new entry that is occurring or is likely to occur in radio-based services.[3] Cable companies are beginning to reconfigure their networks to compete in some fashion with the LECs. PCNs will begin to appear in a few years.

No one can foresee how the local access market will evolve. A few years ago it was widely believed that high-definition television would probably be an analog system. Until just recently, there was considerable doubt that telephone companies could deliver video services without extending fiber optics loops all the way to the subscriber. Given the rapid and unpredictable nature of technological change, regulators should allow relatively free entry and allow the market to determine the best arrangements for connecting residences and business to communications networks.

Of course, facilitating entry into access markets is not that simple. Entry is more likely if the new entrants can connect readily with incumbents, who, in turn, are obviously not eager to cooperate. Regulation of the terms of interconnection is thus inevitable, but on what terms? This leads to all of the problems of open network architecture, unbundling of basic service elements, collocation, and the provision of information on changes in the technical design of the core network. If regulators cannot know the incumbents' costs, how can they know the costs of the basic elements? I offer no magic solution to these problems.

Deregulation of Noncore Services

Among the most contentious issues in telecommunications policy is the development of safeguards to prevent a regulated carrier from cross-subsidizing competitive services from regulated monopoly profits. I do not intend to revisit the history of the FCC's approach to this problem in successive computer decisions, but the problem is pervasive and is likely to get worse.

As new technologies develop, the size of the monopoly "core" that requires regulation shrinks. Competition in intra-local access and transport area (LATA) toll, central-office services, or even local switching cannot be fully effective if the incumbent carrier is regulated and its rivals are not. Yet once the incumbent is allowed to compete freely in some markets while being regulated in upstream services, complaints of cross-subsidy and denial of comparable interconnection will abound. The choice for regulators is either to require incumbents to shed services successively as they become competitive or to deal with the potential problems by shifting to price caps and enforcing rules of comparable interconnection. Neither "safeguard" is perfect, but the choice should be viewed as just one difficult step along the road to a competitive pluralistic network.

The substitution of price-cap regulation for the current cost-based approach favored by most state regulators will not only reduce the incentives for cross-subsidization by the regulated carriers but also will eliminate a powerful disincentive for technical progress. Cost-based regulation stunts the carriers' incentives to seek cost-saving or revenue-enhancing technologies, particularly when such regulation is accompanied by unrealistically slow depreciation schedules. Even though price caps are far from foolproof, any movement away from cost-based regulation should be eagerly supported by those who want to encourage major network investments by regulated carriers.

Forbearance from Mandating Network Technology

Historically, regulators have assumed a planning role in an attempt to prevent regulated carriers from padding the rate base under rate-of-return regulation. Investments are permitted if the new capital is "used and useful." In some instances, regulators have forced changes in technology to accommodate new services or new competitors, but they have generally stopped short of forcing regulated firms to push out the envelope of technology for fear that such investment will expose ratepayers to too much risk.

Some industry observers and users of advanced services now want to use regulation as a prod to investment in an advanced national information infrastructure or "superhighway." Regulated firms would be given assurances on rates for basic services in return for their commitment to build a high-speed switched network using fiber optics all the way to the subscriber or to the "pedestal." The precise configuration of such a network—whether it employs synchronous or asynchronous technology, for instance—is subject to considerable debate and uncertainty. Nor is it clear which services will be demanded by most subscribers or who will be allowed to provide them.

Given the speed of technological progress in this area, it is particularly dangerous to suggest that regulated firms, operating under a constitutional guarantee that they can recover their investment through subsequent rate increases, should be instructed by regulators to invest billions of dollars in new technology that may either prove unwanted or be surpassed by other technologies. Public utilities are not particularly good in making such decisions, as our recent sorry experience with nuclear power should confirm. Two years after being liberated from the Modified Final Judgment restrictions on information services, the Bell operating companies have not been very successful in developing and marketing new content-based services.

There is another important reason for not forcing (or subsidizing) investment by regulated carriers in advanced networks. Such a policy will only increase the pressure on regulators and legislators to limit competition since these massive new investments must be recovered through ratepayers or taxpayers. This will only increase the argument for protecting cable television companies, newspapers, and other information providers from the potential competition of regulated telephone companies. Rather than expanding the domain of regulated competition, we should be contracting it as rapidly as possible.

Given the bewildering array of technical possibilities for the "network of the future," it would be very dangerous for regulators (or legislators) to exclude competition from certain large players while mandating that telephone companies invest in what is perceived to be today's best choice for the network of the future. At this juncture, it is widely believed that residential demand for video services will drive the network design. But no one can be sure whether video on demand

(VOD) will completely replace today's multichannel offerings or just how it will be delivered. Will paired copper wires suffice for the last mile? Will coaxial cable systems be the best local loop for VOD? Can the current coaxial systems be modified to offer truly universal switched wide-band services, or will such services be offered best by telephone company asynchronous transfer mode (ATM) networks? One shudders at the thought of state regulatory commissions making these decisions.

CONCLUSION

Students of regulation have seen how it is the enemy of technical progress. Regulatory accounting, entry protection, and fears of cross-subsidies often prevent regulated firms and potential entrants from exploiting technological progress. The introduction of new freight cars, piggyback services, hub-spoke airline networks, multichannel urban coaxial cable systems, pay-television channels, and a myriad of telephone terminal equipment was delayed by regulators bent on protecting the public from the evils of monopoly restrictions of output. Given the extraordinary current rate of technical progress in communications, they should be reluctant to limit the domain of current players or of potential entry. Instead, regulators should look to open markets, constrict the core of regulated monopoly, and increase the number of participants in the communications sector.

Of course, a reliance on competition to develop network infrastructure will not please those who think they already know what a modern telecommunications network should look like. I would prefer this risk to repeating the mistakes made earlier in this century by regulating airlines, trucking, broadcasting, cable television, and telephony in the first place.

REFERENCES

Baumol, William J., and David F. Bradford. 1970. "Optimal Departures from Marginal Cost Pricing," *American Economic Review* 60(3):265-283.

Crandall, Robert W. 1991. *After the Breakup: U.S. Telecommunications in a More Competitive Era.* Brookings Institution, Washington, D.C.

Faulhaber, Gerald R. 1975. "Cross-subsidization: Pricing in Public Enterprises," *American Economic Review* 65(5):966-977.

Huber, Peter W., Michael K. Kellogg, and John Thorne. 1992. *The Geodesic Network II: 1993 Report on Competition in the Telephone Industry.* Geodesic Company, Washington, D.C.

Noll, Roger G. 1989. "Telecommunications Regulation in the 1990s," *New Directions in Telecommunications Policy,* Paul R. Newberg, ed. Duke University Press, Durham, N.C.

Palmer, Karen. 1992. "A Test for Cross Subsidies in Local Telephone Rates: Do Business Customers Subsidize Residential Customers?" *The RAND Journal of Economics* (Autumn):415-431.

Perl, Lewis. 1985. "Social Welfare and Distributional Consequences of Cost-based Telephone Pricing," paper presented at the Annual Telecommunications Policy Conference, Airlie, Virginia.

Taylor, William E., and Lester D. Taylor. 1993. "Postdivestiture Long-Distance Competition in the United States," *American Economic Review: Papers and Proceedings* 83(2):185-190.

Winston, Clifford. 1993. "Economic Deregulation: Days of Reckoning for Microeconomists," *Journal of Economic Literature* 31(3):1263-1289.

NOTES

1. Roger Noll has called this approach to regulation "residual regulation." See Noll (1989).
2. Huber asserts that competition is now more likely in local access/exchange services than in long distance.
3. Nextel is already operating as a third cellular service.

State Regulatory Policies and the Telecommunications/Information Infrastructure

Robert G. Harris

INTRODUCTION

Historically, the term "infrastructure" has been applied to public-sector investments in highways, airways, schools, and libraries. Because virtually all investment in the communications infrastructure was done by private business enterprises, there was a tendency to take it for granted. No longer. Historically, the United States has relied on a dual system of regulation, federal and state, that strictly limited competition and provided for a nearly guaranteed rate of return to private investors, to ensure the development of a ubiquitous public switched telephone network with affordable rates. It has become evident that the traditional regulatory policies for achieving and sustaining universal service are no longer viable. Because of dual jurisdiction, though, there has been a growing gap between procompetitive federal regulatory policies and those of many states, which are clinging to the regulatory regime of the past.

In this paper I briefly review some of the changes that have been and will be occurring in telecommunications technology, markets, and public policies. I explain the implications of these changes for the telecommunications/information infrastructure in the context of global competition and the strategic character of telecommunications and information services. I consider the implications of the changing character of telecommunications infrastructure for public policy objectives, emphasizing the need for public policies that place greater emphasis on dynamics than statistics, on the grounds that innovation and productivity improvements will generate more benefits than efforts to control prices or limit competition in hopes of maintaining historical cross-subsidies. I also briefly describe several recent state regulatory policy changes or pending initiatives that are consistent with the following progressive policy objectives: reducing rate distortions in the pricing of telecommunications services; reducing entry barriers in local exchange services; deregulating competitive or discretionary telecommunications services; and the adoption of price regulation, rather than rate-of-return regulation, of noncompetitive telecommunications services.

CHANGES IN THE TELECOMMUNICATIONS INDUSTRY

In the past decade or so the United States and other highly developed economies have entered the postindustrial era. In the industrial age the extraction of natural resources for energy and raw materials and the manufacturing of goods were the chief drivers of economic growth. While manufacturing continues to be important, employment in the service sector continues to grow, owing mainly to the tremendous advances in computers and communications. In the past century, agricultural employment has declined from 45 percent to less than 5 percent and employment in

125

manufacturing has returned to its 1890 level of 20 percent after peaking at 30 percent in 1960, while employment in services has exploded from 30 percent to over 77 percent (and 75 percent of the gross national product). Even in manufacturing industries, knowledge-based service activities (e.g., information processing, communications, research and development) constitute 65 to 75 percent of manufacturing costs and even more of the "value added" in the manufacturing sector (Quinn, 1992, pp. 3-30).

What railways, waterways, and highways were and are to the *goods* economy, telecommunications networks are to the *service* economy, since a very large share of value creation in the service sector involves the generation, manipulation, storage, retrieval, and other use of information. Today, information-based enhancements have become the main avenue to revitalize mature businesses and transform them into new ones. As chronicled by George Gilder in *Microcosm*, the basis of this transformation is microelectronics technologies and their application to computers, communications, manufacturing equipment, consumer products such as autos and household appliances, and virtually all service industries (Gilder, 1989, pp. 317-383).

As Davis and Davidson (1991) have pointed out, "Today, *information-based enhancements have become the main avenue to revitalize mature businesses and transform them into new ones.* In every economy, the core technology becomes the basis for revitalization and growth. Information technologies are the core for today's economy, and to survive all businesses must *informationalize*" (p. 17; emphasis in original).

To understand the role of telecommunications in that transformation, one must consider three major types of change in telecommunications—namely, changes in technology, markets, and public policies. In combination with the relatively faster growth of service industries, these changes have heightened the importance of telecommunications in economic development and, therefore, increased the value of *developmental* telecommunications policies (i.e., policies that promote, rather than inhibit, the development of a vital telecommunications sector).

After several decades of steady, but incremental, technological innovation and adoption in telecommunications, there has been a virtual explosion of new technologies in the past decade. Along with computers, telecommunications is on center stage of the microelectronics revolution: the application of transistors, semiconductors, integrated circuits and other microelectronics in telecommunications equipment has dramatically reduced equipment costs, improved the quality of service, and generated a host of new services and capabilities in the public switched telephone network (PSTN).

Through microelectronics the digitization of telephone switching has made possible many new services and reduced the costs of enhanced services. Digitization and optical technology in interexchange transmission, interoffice trunking, and cable TV distribution systems have reduced the costs of those services and created entry opportunities for cable companies and alternate access providers such as Metropolitan Fiber Systems. As federal policies and market competition have driven the prices of interexchange and data communications services toward costs, technological change has induced substantial increases in demand, thereby inducing investments in capital-embodied technological innovation.

At the same time, these developments have eroded the traditional natural monopoly character of the PSTN and stimulated the rapid market penetration of private branch exchanges, private telecommunications networks, facilities-based carriers, and resellers. Thus, local exchange carriers face competition from equipment suppliers (e.g., Centrex services competing with PBX vendors); with their own customers, who increasingly turn to self-supply of switching and network services; with competitive access providers for local exchange services; with established carriers like AT&T, MCI, and Sprint for switched and dedicated services; and with private pay phone vendors for public phone services.[1]

Contemporaneous with these changes in the wireline telephone network, technological developments in radio communications (including microwave), satellite (including direct broadcast satellite to customers' premises), terrestrial broadcast radio and television, and cellular telephony have dramatically lowered the cost, improved the quality, and stimulated the proliferation of a wide range of wireless communications services. Not only has competition grown rapidly in each of these communications media, but as the capacity and range of services available through each medium have increased, so too has the competition among communications media. So, for example, sideband FM and TV broadcasting are now used to distribute financial market data and credit card verification information, in direct competition with the PSTN.[2]

There is every reason to believe that rapid technological change in both wireline and wireless telecommunications will continue, at accelerating rates, into the indefinite future. One major boost to innovation is the growing quest for technological advantage by leading competitor nations to the United States. Once the unrivaled leader in telecommunications equipment technology and system deployment, the United States faces competition from equipment manufacturers from Japan, Germany, France, and the United Kingdom, among others. Companies from these countries have committed major resources to telecommunications research and development and have caught up with U.S. suppliers in some technologies and passed them in others.[3] This heightened international competition will almost certainly accelerate technological change, increasing the risk of economic obsolescence in the PSTN, but also expanding the potential benefits of investment in new technologies.

Dramatic changes are also occurring on both the supply side and the demand side of telecommunications markets. As little as ten years ago, local exchange carriers faced very limited competition for local network access. Today, traditional local telephone carriers face competition from a host of competitors and potential competitors: competitive access providers, interexchange carriers, cellular carriers, cable TV carriers, and, soon, personal communications services providers. Even small businesses and residential users have an increasing array of alternatives to PSTN services. For example, one can now use, in certain localities, (1) FM sideband for delivery of stocks and bonds price information to personal computers; (2) public TV sideband for delivery of credit card validation information to retail point-of-sale terminals; (3) a cable TV carrier for provision of local service bypass, local area networks, and metropolitan area networks; (4) cellular mobile as a substitute for coin-operated telephone service; and (5) a CD-ROM or magnetic disk database as a substitute for on-line access to information. More generally, there is growing competition between PSTN service providers and equipment vendors, including "SmartSets" (competing with "custom calling features"), auxiliary equipment such as answering machines (competing with voice mail), and personal computers.

Recent policy decisions by the Federal Communications Commission (FCC) to further open up local access to competition portend even more rapid change.[4] The allocation of spectrum to facilitate the development of personal communications networks is one of the last nails in the coffin of the local exchange monopoly. Not surprisingly, other telecommunications companies are rushing to exploit new technological and regulatory opportunities in local exchange services. Recent announcements of the formation of strategic alliances among interexchange carriers, cellular carriers, and cable television carriers, such as AT&T's acquisition of a 33 percent interest in McCaw Cellular Communications (Karpinski, 1992a) and the Time Warner-MCI switched access trial in New York City (Karpinski, 1992b), will accelerate the emergence of vigorous competition for local exchange services.

The dramatic changes on the "supply side" of telecommunications markets have been matched by equally significant changes on the "demand" side. Rapid growth in the use of computers, data, and transactions processing systems (e.g., electronic funds transfers, credit card verification, automated teller machine networks, travel reservation services) has induced demand for data communications services, which is growing much faster than voice communications. With recent developments in computer graphics and image processing and storage systems, it is becoming evident that data will be superseded in the near future by images as the fastest-growing share of communications traffic (e.g., American Express used to first keypunch data from credit card transactions, then move those *data* electronically from place to place; it now takes a "picture" of the credit receipt and moves the *image* from place to place). These changes from voice to data and image communications explain the need for and use of broadband transmission media.

As the demand for sophisticated telecommunications applications has grown, large business users have developed specialists in managing and purchasing telecommunications services. In just the past ten years, more than half of the Fortune 500 and thousands of medium and smaller enterprises have created a "chief information officer" position, to whom a range of computer, communications, and information experts and analysts report. With intimate knowledge of the technical and economic alternatives, these buyers continually seek out and exploit small differences in prices and have the capacity to assemble integrated systems from purchased "piece-parts." This means, in turn, that when regulated prices vary from market realities, buyers will turn to more economic alternatives.

Even among residential users, there are rapidly growing demands for advanced telecommunications and information services. With nearly 30 percent of the U.S. population engaged in work at home, and with more than 35 million personal computers in American homes, it is simply no longer true that residential customers will be satisfied with "plain old telephone service."

The divestiture of the Bell operating companies by AT&T and competition policy decisions by the FCC, in combination with the changes in telecommunications technology and market conditions, have facilitated competition in customer premises equipment, interexchange services, and, most recently, local exchange services. The FCC has consistently pursued procompetitive policies for the past decade, often in opposition to states, sometimes requiring the use of preemption over state regulations (e.g., customer premise equipment, inside wiring decisions). Clearly, the general direction of federal policy is procompetitive (NTIA, 1988).

ROLE OF THE TELECOMMUNICATIONS INFRASTRUCTURE

Some states have begun to liberalize their telecommunications regulatory policies since divestiture in recognition that traditional policies cannot cope with rapid changes in technology, competition, and market conditions. States such as Michigan, New Jersey, and Tennessee are explicitly using reformed telecommunications policies to promote state economic development, improve public services, expand educational and social services, and create a better economic environment to attract businesses and skilled jobs. As more and more states implement progressive, procompetitive policies in telecommunications, the cost of not keeping pace with these changes, in terms of lost jobs and economic development to other states, will increase.

These recent changes in state policies and perspectives reflect a growing awareness of the vital role of telecommunications in economic development. There is, in fact, a deep intellectual tradition that views telecommunications and other infrastructure industries in this way. In his seminal work, *The Strategy of Economic Development*, Albert O. Hirschman introduced the notion of backward and forward linkages from an infrastructure sector (e.g., energy, transportation, communications) to supplier and user industries (Hirschman, 1958). Hirschman explained why and how high investment in sectors with strong linkages will lead to more rapid economic growth. He coined the term "social overhead capital" to describe infrastructure industries that:

- Provide services that are basic to a great variety of economic activities;
- Exhibit a high degree of "publicness" (and are therefore usually provided by public agencies or private firms under public control);
- Are immobile and therefore cannot be imported;
- Have substantial "lumpiness" or technical indivisibilities; and

• Have very high capital/output ratios, with large fixed investment required to achieve an economically viable level of output.

Given these characteristics, Hirschman argued that investment in social overhead capital not only was essential, but also required a national strategy for economic development. Hirschman's ideas provide the intellectual underpinning of national and state investments in railroads, waterways, highways, airports and air traffic control systems, hydroelectric projects, and public schools and universities.[5] None of these investments could have been justified solely on the basis of a private cost-benefit calculation. Initially, both the Interstate Highway System and the air transportation system were used by a relatively small share of the population, often traveling for business purposes. Yet everyone contributed to the cost of the public investments through:

• Preferential tax treatment (e.g., exemption from property taxes and tax-free financing of public debt);
• Collection of excise taxes on gasoline used on city streets (the construction, maintenance, and operation of which are paid for by property taxes) to finance construction of intercity highways; and
• Deliberate cross-subsidies (e.g., distribution of gasoline excise tax revenues from urban to rural areas).

The justification for public expenditures on these infrastructure systems was based on the "positive externalities" and the economic development they generated. That is, beyond the benefits realized by direct users, these investments contributed substantially to economic growth and development in the local, regional, and national economies. Investment in infrastructure can be justified by the fact that what consumers have to *spend* on goods and services depends on what they *earn* as workers. It is inconceivable that U.S. workers could have experienced the same growth rates in productivity, income, and standard of living in the postwar period without these infrastructure investments. By the same token, the recent slowdown in productivity growth and falling real incomes reflect, in part, declining infrastructure investment relative to our competitor nations.

Although the telecommunications infrastructure has been mostly privately owned and operated, telecommunications policy in the United States was, for fifty years, based on the same basic principles as public-owned infrastructure. In lieu of public investment, the United States used industry regulation to reduce the risk of, and ensure reasonable returns on, telecommunications infrastructure investment in order to induce sufficient investment of private capital. The goal of universal service was based on a recognition of the positive externalities of additional customers on the network and the stimulus to economic development.

Even though we have achieved universal service, we should still remember the principle of economic development on which it was based. In fact, the principles of universal service, positive externalities, and social overhead capital are as applicable to the next generation of telecommunications services as they were to the last. Much of the value of integrated digital services to any given user, for example, will depend on how many others have access to and are interconnected with them. These externalities are especially powerful in complementary industries, such as information services. Capabilities built into the public switched network could greatly facilitate access to, and thereby demand for, enhanced information services. This is especially true for small-business and residential users, who have neither the resources nor the expertise to design, build, and operate their own private communications system, as do large corporate and governmental users.

In addition to the positive externalities, infrastructure investment in telecommunications improves the quality of services, increases the number of services, and reduces the costs of those services. Thus, telecommunications investment generates substantial benefits to the users of telecommunications services. In the past decade or so, telecommunications services have literally revolutionized many industries; in virtually all industries and sectors, telecommunications services have generated or have demonstrated the potential to generate major productivity gains. Five recent studies of economic development support my conclusion that telecommunications services are playing an increasingly important role, especially in the high-technology, knowledge-intensive industries that generate skilled jobs and a high degree of learning on the job.

First, a study by the Organization for Economic Cooperation and Development (1988) found extraordinarily rapid growth in the importance of telecommunications services to business users in all member nations, with crucial effects on international competition in telecommunications-intensive industries, especially financial services (banking, insurance and securities, commodities, and foreign exchange trading), publishing and information services, wholesaling and retailing.

Second, in a study of the impact of information technologies on service industries (financial services, health care, insurance, and publishing), James Brian Quinn (1987) found:

- Substantial forward linkage economies and externalities, including realization of economies of scale and economies of scope (the capacity to provide entirely new service products through the same service network);
- A substantial increase in "output complexity" (the quantity and quality of services available to customers);
- A blurring of industry boundaries through functional cross-competition; and
- Improved international competitiveness, through the locational decisions of manufacturers who use these services.

One of the most valuable "downstream" benefits of telecommunications services that Quinn found is the increased geographic extensiveness of user industries. In urban areas this extension improves accessibility and enhances competition among providers of goods and services (e.g., automated teller machines competing with branch banks; telemarketers competing with local retailers). In rural areas the geographic extension of services through telecommunications often means a substantial improvement in the quality and variety of services available to rural consumers and businesses, or even the difference between having affordable access to a service or not (e.g., remote health care services).

In a recent extension of his work, Quinn found that telecommunications can make significant contributions to increased productivity and improved competitiveness in manufacturing as well. In *Intelligent Enterprise*, Quinn (1992) explores the revolutionary changes in organizational and industry structure that are being driven by the application of knowledge and information, noting that:

[d]iscussions concerning America's manufacturing competitiveness have consistently overlooked an area that offers major productivity leveraging possibilities: the manufacturing-services interface. On the one hand, service companies have become some of the most important customers, suppliers, and coalition partners for many manufacturing concerns. U.S. service enterprises are both near at hand and are among the world's most efficient performers—surpassing the services productivity of virtually all other advanced industrial economies, especially Japan. Major opportunities exist for manufacturers to exploit U.S. service companies as major customers, as lead companies or co-developers for new products, as potential suppliers, as value-adding advisers or market intermediaries, and as sources of valuable information and distribution clout in their markets. (p. 208)

In order to realize this potential, Quinn urges increased investment in communications infrastructures and regulations that are "goal-oriented rather than means-specifying" (Quinn, 1992, pp. 432-433). By "goal-oriented," Quinn means regulations that use incentives to stimulate desired performance, rather than attempting to mandate the means of achieving such performance through "command-control" regulations.

Third, in a study commissioned by the state of New York, Coopers and Lybrand found that purchases of telecommunications services by businesses in the United States were growing at the rate of 11.8 percent per year, compared with gross national product growth rates averaging 2.5 percent (Coopers and Lybrand, 1987). Employment is growing fastest in "telecom-intensive" industries. Even though the real prices of telecommunications services have been declining, purchases of telecommunications services are a growing share of business costs. Consequently, business users are more price sensitive and mobile, so the local cost and quality of telecommunications services play an increasingly important role in business location decisions. Telecommunications can also geographically extend the workplace; by so doing, telecommuting can enhance the quality of workers' and families' lives, bring economic opportunities to rural areas, and aid in reducing congestion and air pollution in urban areas.

Fourth, in a study commissioned by the New Jersey Board of Public Utilities, Deloitte and Touche (1991) found that:

- "As New Jersey continues to move toward an information services-based economy, today's local exchange carrier network will increasingly constrain users' (especially residential and small business users) ability to fully participate in the Information Age";

- "A significant opportunity exists to advance the public agenda for excellence in education through improvements to the telecommunications infrastructure";

- "Strong motivation, especially in the improved quality of care and cost reduction, exists for increasing the use of telecommunications and information technologies in the health care industry in New Jersey"; and

- "Public policies that encourage deployment of an advanced telecommunications infrastructure are essential for New Jersey to achieve the level of employment and job creation expected in the state." (p. I-3)

Fifth, on the basis of an extensive survey and focus group interviews of small businesses, the Illinois Task Force on Advanced Telecommunications and Networking (1992) concluded that:

- "All enterprises are becoming more information intensive and . . . [the] Illinois communications and computing infrastructure will define state economic development capabilities in the future"; and

- "A robust telecommunications infrastructure is vital to meet the requirements of education and training, libraries, health services, safety, and other social components which collectively create the quality of life." (p. 9)

Each of these studies has confirmed that the direct user benefits, positive externalities, and economic development benefits of telecommunications are, if anything, growing over time. For an increasing number of industries, access to advanced telecommunications services will be essential to competitive advantage, possibly even competitive survival, in global markets. Until recently, though, little or no thought was given to the implications of U.S. telecommunications policy for international competition, perhaps because the United States was a hegemonic economic power in the world. That extremely parochial view of the world is increasingly out of touch with current economic and technological reality, especially considering what has happened and is happening elsewhere.

In an increasing number of countries, national policies are premised on the idea that telecommunications is a strategic industry, with economic characteristics that require or deserve special consideration in assessing policy alternatives. While the term "strategic industry" is often used indiscriminately (i.e., meaning that an industry is *important*), there is a rapidly growing literature analyzing the economic conditions and characteristics of an industry that make it strategic. Although that literature is concerned primarily with international trade theory and policy, and hence addresses "tradable goods," much of the developing theory applies to nontradable goods as well and is therefore relevant to telecommunications services (see, e.g., Krugman, 1987; Porter, 1990).

IMPLICATIONS FOR REGULATORY POLICY OBJECTIVES

The central message of these international developments is that state and federal telecommunications policies must, in order for the United States to be internationally competitive in the information age, explicitly acknowledge and incorporate the broader economic developmental benefits of telecommunications services. The United States can no longer regulate telecommunications as if it were a purely domestic industry, when other nations have chosen telecommunications as a leading instrument of national economic development and competitive advantage.

Although the telephone regulatory process has become quite adversarial in recent years, historically there was a very strong social and political consensus regarding the goals and objectives of telephone regulation. Moreover, there was a consensus among policymakers that rate-of-return regulation was the best means of achieving those goals. As the economy becomes increasingly dependent on telecommunications services, though, the consensus over both ends and means has been breaking down, as those involved in the policy process attempt to cope with the revolutionary changes that have occurred, and will continue to occur, in telecommunications.

Changes in telecommunications technology and markets have two crucial implications for public policy. First, the goals and objectives, the *ends*, of regulatory policy should reflect the growing importance of telecommunications to the economic welfare of households, businesses, and public agencies. This requires that policymakers give added weight to economic development effects in considering policy alternatives. Second, in developing and implementing a regulatory framework, the *means* to achieve those policy objectives, policymakers should take full account of the dynamics of change in telecommunications, on both the supply and the demand sides. Policies that worked well in the past, in a markedly different industry environment, will not work well in the current and future environments of telecommunications.

Traditional Regulatory Policy Objectives
in Telecommunications

The era of stable rate-of-return regulatory policy in the telephone industry spans the period from the 1920s through the early 1970s. The dual system of federal and state regulation was reasonably successful, in large part because there was a strong consensus on policy objectives, which included the following:

- The primary objective of telephone regulation was, historically, the achievement of *universal service*, that is, promoting the development of a ubiquitous, affordable telephone network, with prices that account for need (social equity) as well as ability to pay (market). This goal was typically reflected in telephone rate designs that priced some services, such as toll, business services, and custom calling features, above cost in order to keep the price of basic residential service at affordable levels. The subsidy to residential users generally included a preferential price for access and local usage (in many states, flat-rate local calling). The universal service objective was also the justification for geographic price averaging, by which customers in rural areas paid the same rates for service as customers in urban areas, even though the cost of both local and toll service was significantly higher in rural areas.

• A second important objective of regulatory policy has been maintaining a consistently high level of *service quality*, including timely installation of new service; very low "call blocking" rates; very little downtime in the network or subscriber lines; and the ability of the public switched network to withstand or quickly recover from natural disasters and other crises. In unregulated industries, market forces generate a wide range of prices and product qualities. In regulated industries, such as electricity and telephone, the systems were engineered to very high standards, whether or not the revenues from any given class of customers cover the cost of the services they use.

• An often implicit, but nonetheless important, third policy objective has been *rate stability*, that is, minimizing changes in rates for a class of services or a class of customers. In the rate design process, substantial weight has been given to the existing rates in order to protect customers from "rate shock" due to sudden, unforeseeable price changes.

• Providing investors with a *fair rate of return* has been a fourth traditional policy objective of telephone regulation. Fair rate of return encompasses two purposes: (1) giving investors fair compensation for the risk and time value of capital and thereby (2) stimulating private investment in "social overhead capital," the telephone infrastructure. These twin purposes have been served by limiting competition to reduce economic risk (by granting exclusive franchises) and by reducing financial risk through a steady, predictable rate of return on investment.

Progressive Regulatory Policy Objectives

Even accepting that universal service and other social equity goals must continue to be taken into account, the objectives of telecommunications policy should be changed in recognition that economic efficiency, infrastructure investment, and modernization and technological innovations in telecommunications are increasingly important to the productivity of private enterprises, not-for-profits, and public agencies and, therefore, to the competitiveness of the state economy. To take account of the economic development and productivity-enhancing potential of telecommunications, public policy should explicitly incorporate, and give greater weight to, the following objectives:

• One only need consider the magnitude of the U.S. trade deficit and federal and state fiscal deficits to understand why *efficiency* should be given more weight in public policy decisions. Because the United States is no longer a hegemonic economic power in the world, and because national productivity gains have not kept pace with leading competitor nations, the country can no longer afford the "luxury" of public policies that fail to promote economic efficiency objectives. Either policymakers elevate efficiency goals to get better use of and more from the nation's economic resources or they will have to do with fewer resources to meet economic and social objectives. That means (1) either allowing market forces to set prices or, if regulated, allowing prices to reflect market factors to achieve allocative efficiency; (2) using incentives to promote technical efficiency by telecommunications companies and their employees; and (3) reducing unnecessary regulations to achieve administrative efficiencies in the administrative process.

• In a predominantly market economy, public policies should attempt to be *responsive to market conditions* in the industry being regulated and in related industries. Prices, as signals of cost and value, play a critical role in market exchange; therefore, regulators should allow prices to be set by market forces whenever possible or, alternatively, emulate market prices when they do set prices (or pricing bands or rules). Similarly, regulators should, to the maximum extent possible, allow markets to determine what variety of products and services will be offered. Market pressures have increased the rewards of good public policies (i.e., those that stimulate investment, increase usage, and promote economic development in the states). Market pressures have also increased the costs of policies that are not consistent with market conditions (e.g., uneconomic bypass, self-supply, relocation of facilities).

• Technological *innovation* is a third critical policy objective for promoting economic development. By facilitating innovation and the adoption of better technologies, policy-makers can ensure that users will obtain the benefits of innovation, especially lower costs through productivity gains. Regulators can also promote innovation by enabling the rapid introduction of new services to meet customer needs and by allowing greater pricing flexibility for discretionary services.

• Historically, U.S. policies toward public infrastructure industries have incorporated, if only implicitly, *economic development* goals. Accordingly, the United States has been an international leader in public infrastructure investments, which have played a significant role in the nation's economic growth and global leadership over the past several decades. Unfortunately, public recognition of similar economic benefits from private infrastructure investment has lagged. Only recently, as other nations have turned their attention to telecommunications services as a means of promoting economic growth, as an instrument of national economic strategy, have U.S. policymakers begun to take sufficient notice of the positive externalities and other public benefits of telecommunications. Elevating the importance of infrastructure investment among policy criteria can potentially improve the performance of telecommunications industries and the nation's economy.

• One of the most important steps toward efficiency objectives has been taken by allowing, even encouraging, *competition* in telecommunications services and equipment. Strictly speaking, competition is not a policy objective but a means for achieving other objectives. Still, because there is such a strong consensus in the United States that competition is desirable, policymakers have often elevated competition to a goal of regulation. Competition is especially effective in promoting efficiency, since market competition drives prices toward costs and causes suppliers to minimize costs. Ultimately, competition can supplant or even eliminate the need for regulation, although in the meantime quasi or partial competition may well make the task of regulators more difficult.

Balancing Competing Objectives

Good public policy decisions require making trade-offs among competing objectives. There will be times when some policy objectives conflict with others; good policy recognizes and balances among multiple policy objectives. Policymakers should not assume these trade-offs are inevitable in all cases, however. The art of designing outstanding policies involves finding ways to reduce the trade-off, to get *more* equity and *more* efficiency. Properly designed and implemented, policies can reduce trade-offs among competing objectives; the resulting improvements in performance can generate increased efficiency and also pay a consumer "dividend."

IMPLICATIONS FOR STATE REGULATORY POLICIES

Until the 1980s there were few significant differences in state regulatory policies toward common carrier telephone companies. In the past decade, though, the states have taken very different paths in their respective responses to the AT&T divestiture, changes in federal regulatory policies, and a growing recognition of the importance of telecommunications to economic development. From a national policy perspective, this variation among the states is useful in generating experience with alternative regulatory policies—a classic case of "laboratory experiments among the states." It is my position, though, that there are many cases in which state policies are inhibiting the growth and development of the telecommunications infrastructure. States should be encouraged, if not required, to adopt policies that meet certain minimum requirements. In this section I discuss four types of regulatory reforms that some states have adopted or are in the process of adopting.

Reducing Rate Distortions:
California's Implementation Rate Design

In Phase 3 of its New Regulatory Framework, the California Public Utility Commission (PUC) has expressed its intent to open its local access and transport areas (LATAs) to competition, with 10xxx dialing, for local, extended area service, zone usage measurement, toll, and 0+ calls. The decision, now pending, would raise the price of flat-rate residential service from $8.35 to $13.00 per month, while maintaining a 50 percent discount for lifeline customers. It would substantially reduce the toll rates of Pacific Bell, the major local exchange carrier in the state: in the 17- to 20-mile band, from 22¢ to 9.75¢/minute and in the 51- to 70-mile band, from 37¢ to 13¢/minute. It would also simplify and expand the applicability of usage discounts, making it easier for customers to choose a rate plan that would minimize the prices they pay for toll services.

Through this decision, the California PUC is finally accepting competition as an essential element of the state's telecommunications policy. By removing a large share of rate distortions—that is, prices that are either significantly above or below economic costs—the reformed policy will increase efficiency, promote economic entry by competitors, and encourage the adoption of information technologies that are complementary to communications usage. Still, the PUC proposal leaves substantially unresolved the following problems: continuation of flat rate pricing (i.e., free local calling for residential subscribers) distorts choices between wireline and wireless technologies. Although reduced in magnitude, continuing residential rate subsidies distort the customer's choice between analog and digital service (i.e., the price differential between subsidized analog service and unsubsidized integrated services digital network (ISDN) service biases customers from adopting the advanced service). The plan maintains price averaging, which distorts the economics of rural service and deters the entry of lower-cost competitors and the development of lower-cost technologies.

So, while the California plan represents a significant step in the right direction, it came only after years of highly contentious proceedings and has still not yet been officially adopted. Moreover, given the rate at which technologies are changing and market competition is emerging, it is apparent that regulators, even if heading in the right direction, have difficulty keeping up with, much less getting ahead, of industry development.

Removing Entry Barriers to Local Exchange Competition

Ameritech has filed a joint state and federal regional plan that offers a quid pro quo: Ameritech would unbundle its local network and open the local exchange market to greater competition, in return for being allowed into the interexchange market. The plan would require physical interconnection and collocation of competitors into Ameritech's central offices with reciprocal traffic arrangements. It would also unbundle all critical network elements: local loop, switching, transport, SS7 call setup, and numbering. It would offer customers a choice of carriers, with usage presubscription, and, as soon as technology allows, 1+ dialing for all toll calls. The plan also would shift the universal service burden to all competitors through a "bulk billing" arrangement.

To be implemented, the Ameritech plan requires the approval of the FCC, the U.S. Department of Justice, and the District Court, as well as the regulatory commissions of the five states in which Ameritech operates as a local exchange carrier. The proposal illustrates the high degree of interdependence between federal and state regulatory policies and between antitrust and regulatory policies. It also demonstrates that timing in regulatory reform is often critical: Ameritech seeks "simultaneous" implementation of opening local markets and allowing it into inter-LATA services, while competitors take the position that Ameritech should unbundle and open the local exchange market first, and then, after a sufficiently large loss of market share, Ameritech could be allowed in interexchange services.

Regardless of whether the Ameritech plan is approved as proposed, it is evident that variations in state policies toward competition are growing. It raises the question of whether the United States can continue to tolerate such large differences in policies that are so fundamentally important to the national interest in telecommunications infrastructure. Pending legislation in Congress could preempt states whose policies prohibit competition in local exchange services. Given the number of states that continue to oppose competition, it may well be in the national interest to adopt such preemptive policy.

Deregulating Noncore Services and
Regulating the Price of Core Services:
Price Cap Plans

Six states have adopted price caps or rate freezes with no earnings sharing as part of their incentive regulation plans. All of these plans have several key elements in common: they deregulate or detariff services that are competitive or discretionary; they remove limits on the rate of return earned by the local exchange carrier; and they institute price freezes or price indexing for core or essential services. Summaries of each of these plans follow.

Kansas

The main features of TeleKansas, Southwestern Bell's incentive regulation plan, are as follows:[6]

- The plan froze basic local residential and business rates for the period 1990 to 1995.
- Local rates could increase for some customers due to the elimination of some party lines and exchange reclassifications.
- Bell's long-distance rates were permanently reduced by $17.1 million.
- Access charges for long-distance calls were cut by $2 million.

• Bell customers will save about $21.3 million in each of the first two years of the plan and $22.8 million in the remaining three years, for a total of $110 million.

• Bell committed to an investment of $160 million over the next five years to modernize telephone facilities, upgrade all customers to one-party service, and eliminate all 911 basic service charges.

• The plan changed the way Bell is regulated (it removed the regulatory cap on Bell's earnings).

• Pricing flexibility on certain discretionary products was allowed by shortening the approval process to 20 days from 30 days.

Since its adoption, TeleKansas has resulted in increased infrastructure investment, more new services to justify financially those investments, and rate stability to local service customers. The telephone company had impressive first-year revenue increases while maintaining rate stability and increasing infrastructure investment.

Michigan

Michigan's Telecommunications Act of 1991 essentially replaced rate-of-return regulation with service-by-service regulation. The main features of the plan are the following:

• Monthly service rates were frozen for all but the smallest local exchange carriers for two years.

• Residential rates were flat rated up to 400 calls.

• Basic local exchange rates may not cross-subsidize other services.

• Access rates continue to be regulated and are capped at interstate rates unless approved by the FCC or upon agreement of the parties.

The deregulation law was passed too recently to draw many conclusions, but no basic service rate increases have occurred since the plan was adopted. There have been some long-distance rate increases.

Nebraska

LB 835 deregulated all telecommunications company rates with one exception: basic local exchange service. Tariffs introducing new services or altering rates for existing services can be implemented 10 days after they are filed at the FCC. Basic local exchange service rates are not subject to traditional rate-of-return regulation. Basic local exchange rates are covered under various provisions of the law: (1) The telephone companies must give their customers 120 days' notice and hold a public informational meeting in each commission district before they can change their rates; (2) consumers can authorize the Public Service Commission (PSC) to review a rate increase if enough of the subscribers affected by a rate increase sign a petition; (3) the PSC can review basic local exchange service on its own motion if rates go up by more than 10 percent in a year.[7]

Contrary to claims that telephone rates would double in five years, US West's basic service rates did not increase between 1987 and 1991. Opponents of LB 835 have tried to point out that the constant rates are excessive and have somehow missed a downward trend in national average basic local service rates. Comparison of Nebraska's business and residential rates with those of its neighbors, however, suggests that its rates are not discernibly higher or lower and that they have not missed any "downward trend." In service innovation, LB 835 gave Nebraska an advantage over other states in that it allowed carriers to introduce new services virtually at will. This made it easy for telephone carriers to introduce and test new services (Mueller, 1993, p. 119), which has had a small but measurable effect on investment (p. 138). Deregulation and increased investment are gradually having an impact on Nebraska's technological lead over neighboring states (Mueller, 1993, p. 147).

Vermont

The Vermont Public Service Board approved the Vermont Telecommunications Agreement (VTA) on December 30, 1988. Originally, it was to run for three years but was extended for another two. After a second agreement was withdrawn in 1992, the 1987 social contract was extended through 1997. The main provisions of the VTA were as follows:

- It eliminated rate-of-return regulation for New England Telephone (NET).
- It provided NET with substantial freedom to offer "new services" under rates, terms, and conditions of its own choosing.
- It stabilized local service rates for three years (extended since then).
- It committed over $280 million for network modernization (completed by the end of 1992).
- It maintained quality of service in accordance with specific criteria.

The agreement appears to have had generally positive results:

- Local telephone rates in NET's territory have been stable since 1985. Ninety-five percent of all Vermonters have phones, the fourth-highest rate in the country.
- NET invested $280 million in five years to upgrade its system. The money would have been spent eventually even without the agreement, but more slowly. The new equipment means development officials can assure potential corporate newcomers that they will not be isolated in Vermont.
- More new services have been introduced than in any other state serviced by NET—19 new services.
- The state has spent less time and money regulating NET.
- Regulatory rules have not changed during the agreement, allowing NET, its customers, and its competitors to plan with more certainty. Less regulation means NET and its competitors can focus more on their businesses and customers and less on regulators (Andrews, 1991).
- Infrastructure modernization through the VTA was significant (Edelstein, 1991).

In summary, the VTA did not allow NET to earn excessive profits. On the contrary, the expense of modernization and launching new products resulted in lower than expected earnings for NET. Basic local rates have not increased throughout the life of the agreement, and the telephone network has been modernized significantly.

North Dakota

Some key features of SB2320[8] and SB2440[9] governing incentive regulation were as follows:

- SB2320 classified services as essential or nonessential.
- Rates for essential services were set by the PSC's price factor formula (basically a form of price caps).
- Rates for nonessential services are not regulated.
- SB2440 increased the number of nonessential services and reduced the number of essential services; it also set the price cap in the law.
- Essential service rates have had limited increases allowed through the price cap: the price cap for US West was increased by 1.665 percent in 1992 and by 0.488 percent for 1993; for 1994 the increase will be 0.6 percent.

SB2440 was introduced to clarify and set into the law the list of essential services to be governed by the price cap and to set the price cap and its productivity adjustment. However, rate increases for essential services appear to have been limited. Furthermore, there seems to be broad public support for continuing and expanding the price cap plan, as evidenced by the broad margin of passage of SB2440.

West Virginia

Some key features of the state's incentive regulation plan include the following:

- Effective January 1, 1992, an incentive regulation plan was approved until the end of 1994.
- The plan retained the previous plan's service classification (competitive or discretionary, noncompetitive, and intrastate access services) of an earlier incentive regulation plan and added one more category: services subject to "workable competition."
- Basic service rates were frozen during the plan (a feature of the earlier incentive regulation plan as well).
- Under the incentive plan, C&P agreed to accelerate infrastructure investments toward becoming 100 percent digital by the end of the plan (it was on target to achieve this goal one year earlier, by the end of 1993).

The incentive regulation plan appears to have had a significant impact on the financial health of the telephone company while maintaining rate stability for basic local services. West Virginia's improved telecommunications infrastructure has also attracted telecommunications-intensive firms (Peters, 1991). In summary, its incentive regulation plan appears to have effectively maintained local service rate stability while at the same time providing incentives for C&P West Virginia to rapidly upgrade its network capabilities. Through its Office for the Future program, the state, in partnership with the local telephone company, seems to be successfully marketing those advanced capabilities for outside investors to relocate to the state.

CONCLUSION

Given the rate of change and the magnitude of changes that lie ahead, public policies must place greater emphasis on the dynamics of the industry, whereas regulation has a natural tendency to deal with statics, such as controlling prices and restricting competition. State regulatory policies have an enormous impact on local exchange carriers and the telecommunications/information infrastructure. In too many instances, state regulation has become a major obstacle to competition, deployment of new technologies, and development of new services. The best route to the information infrastructure of the future is not through more regulation but through different and less regulation. Where states are moving in the right direction, they should be encouraged and rewarded. Where they are not, the United States has a strong national interest in limiting the regulatory prerogatives of the states, to ensure regulatory policies that, if not identical, are compatible with national policy goals and objectives.

REFERENCES

Andrews, Richard. 1991. "Telecommunications Agreement Scrutinized," *Vermont Business Magazine,* March 1.

Barr, Francois, and Michael Borrus. 1987. *From Public Access to Private Connections: Network Policy and National Advantage.* Berkeley Roundtable on the International Economy, Working Paper No. 28, University of California, Berkeley.

Coopers and Lybrand. 1987. *State Policy & Telecommunications Economy in New York: Final Report.* New York State Office of Economic Development, Albany, New York.

Davis, Stan, and Bill Davidson. 1991. *2020 Vision.* Simon & Schuster, New York.

Deloitte and Touche. 1991. *New Jersey Telecommunications Infrastructure Study: Report to New Jersey Board of Public Utilities.* Deloitte and Touche, Trenton, New Jersey.

Edelstein, Art. 1991. "Changes Due on Telecommunications Pact," *Vermont Business Magazine,* October 1, p. 27.

Gilder, George. 1989. *Microcosm: The Quantum Revolution in Economics and Technology.* Touchstone/Simon & Schuster, New York.

Harris, Robert G. 1991. "Telecommunications Services as a Strategic Industry: Implications for United States Policy," *Competition and the Regulation of Utilities,* Michael A. Crew, ed. Kluwer Academic Publishers, Boston.

Harris, Robert G. 1993. "R&D Expenditures by the Bell Operating Companies: A Comparative Assessment," pp. 245-266 in *Regulatory Responses to Continuously Changing Industry Structures,* MSU Public Utility Papers, Institute of Public Utilities, East Lansing, Mich.

Hirschman, Albert O. 1958. *The Strategy of Economic Development.* Yale University Press, New Haven, Conn.

Huber, Peter W., Michael K. Kellogg, and John Thorne. 1992. *The Geodesic Network II: 1993 Report on Competition in the Telephone Industry.* Geodesic Company, Washington, D.C.

Illinois Task Force on Advanced Telecommunications and Networking. 1992. *Report of The Illinois Task Force on Advanced Telecommunications and Networking.* Report to Jim Edgar, Governor of Illinois, Springfield, Ill., April.

Karpinski, Richard. 1992a. "AT&T Strides into Local Loop with Possible McCaw Deal," *Telephony,* November 9, pp. 8-9.

Karpinski, Richard. 1992b. "Time Warner, MCI Test CATV Bypass," *Telephony,* December 7, p. 6.

Krugman, Paul. 1987. "Strategic Sectors and International Competition," in *U.S Trade Policies in a Changing World Economy,* Robert M. Stern, ed. MIT Press, Cambridge, Mass.

Mueller, Milton L. 1993. *Telephone Companies in Paradise: A Case Study in Telecommunications Regulation.* Transaction Publishers, New Brunswick, N.J.

National Telecommunications and Information Administration (NTIA). 1988. *NTIA TELECOM 2000.* U.S. Department of Commerce, Washington, D.C.

Noam, Eli M. 1992. "Private Networks and Public Objectives," *Aspen Quarterly* (Winter):106-136.

Organization for Economic Cooperation and Development (OECD). 1988. *The Telecommunications Industry: The Challenges of Structural Change.* OECD Series in Information, Computer and Communications Policy (No. 14), OECD, Paris.

Peters, Eric. 1991. "Telemarketer Opens Bureau in Wheeling," *The State Journal,* December 1.

Porter, Michael. 1990. *The Competitive Advantage of Nations.* Basic Books, New York.

Quinn, James Brian. 1987. "The Impacts of Technology in the Services Sector," in *Technology and Global Industry: Companies and Nations in the World Economy,* Bruce R. Guile and Harvey Brooks, eds. National Academy Press, Washington, D.C.

Quinn, James Brian. 1992. *Intelligent Enterprise: A Knowledge and Service Based Paradigm for Industry.* The Free Press, New York.

Telco Competition Report. 1992. "FCC's Expanded Interconnection Decision Seen as 'Historic' Step Toward Opening Local Exchange Monopoly to Competition," October 15, pp. 1-8, BRP Publications, Washington, D.C.

NOTES

1. For a thorough presentation and analysis of competition in telecommunications, see Huber et al. (1992). For additional evidence on this point, see Barr and Borrus (1987). See also Noam (1992).

2. See Huber et al. (1992), especially Chapters 2 and 4.

3. For a comparison of R&D expenditures by U.S., European, and Japanese telecommunications equipment manufacturers and public telephone operators, see Harris (1993). In a paper presented to the Telecommunications Policy Research Conference in 1989, "Research and Development in Telecommunications—An International Comparison," Thomas Schnoring showed that by 1986, Japanese telecommunications firms had already achieved a higher "technological potential" than U.S. firms overall and had significant leads in optical fiber, image communication, and general picture transmission (Figures 4/1 and 4/2 at pp. 17-18). "Strategic Uses of Regulation: The Case of Line-of-Business Restrictions in the U.S. Communications Industry," by Robert T. Blau and Robert G. Harris, found that among 13 leading global suppliers of telecommunications equipment, the number of electrical patents awarded annually to European and Japanese firms had grown much faster from 1984 through 1989 than those of U.S. firms (Figure 3, p. 183). That study further shows that among the same 13 companies, the "relative technological strength" of the U.S. companies remained about the same from 1984 to 1989, while that of the European and Japanese suppliers grew markedly (Figure 4, p. 184). Finally, the study shows that the global market share of the U.S. companies fell by about half from 1984 through 1989, while the shares of European and Japanese firms grew substantially (Figure 5, p. 184).

4. In reference to the FCC's September 17, 1992, order in Docket 91-141.

5. For an elaboration of this discussion, see Harris (1991).

6. Summarized from "State Modifies Bell Plan to Save Customers $110 Million," United Press International, February 2, 1990.

7. For more detail on the legislation, see Mueller (1993), from which this description was drawn.

8. 1989 *North Dakota Session Laws,* Chapter 566.

9. 1993 *North Dakota Session Laws,* Chapter 469.

The Prospects for Meaningful Competition in Local Telecommunications

Dale N. Hatfield

INTRODUCTION

The purpose of this paper is to explore the prospects for meaningful competition in the provision of local telecommunications services. This topic is important (1) because of its relationship to the line of business restrictions that were placed upon the Bell operating companies (BOCs) as a result of the Modified Final Judgment that produced the breakup of AT&T and (2) because of the efforts of local exchange carriers more generally to be deregulated in the provision of exchange and exchange access services and to expand into information services and other adjacent competitive markets.

PROSPECTS FOR COMPETITION

Cable Company Competition with Telephone Companies

It is useful to look at the public switched telecommunications network (PSTN) as being composed of five parts: (1) the terminal equipment and inside wiring portion on the subscriber's premises; (2) the subscriber access or local loop portion of the network between the subscriber's premises and the local switch or central office; (3) the local switching portion; (4) the metropolitan-area portion of the network interconnecting those central offices and the "points of presence" or switching systems of the long-distance carriers; and (5) the interexchange or long-haul portion of the network. To these five parts—customer premises equipment, local loops, local switching, metropolitan-area/interoffice transport, and the interchange network—it is also useful to add a sixth by dividing the local exchange switching systems into two parts: (1) basic call processing, that is, the lower-level functions associated with making connections between and among subscriber lines and interoffice trunks, and (2) the higher-level, software-based service logic functions and their associated data bases. Of course, it is this separation of basic call processing from the service logic and associated data bases that is leading to the creation of the "intelligent network" and the offering of services like personal number calling.

NOTE: The views and opinions set forth in this paper are those of the author. They do not necessarily reflect the views of the Annenberg Washington Program in Communications Policy Studies of Northwestern University.

At one time—not really so long ago—all of these segments of the network were regarded as being part of one large natural monopoly. The notion that the entire network—all six parts—exhibited natural monopoly characteristics was challenged by (1) rapid technological changes (such as the development of microwave radio) and (2) a lot of rethinking on the part of policy analysts and regulators. There is no need to repeat the regulatory history here, but, briefly stated, the customer premises equipment and long-haul portions of the network were "unbundled" from the four portions of the local exchange network—subscriber access, local switching (2 parts), and local transport. Competition was then allowed in these unbundled portions, and that competition is now well established in both areas. It is probably fair to say that many policy analysts, initially at least, believed that the local exchange network—composed of the four remaining parts—still constituted a natural monopoly.

Viewed in a historical context like this, it is not really too surprising that the natural monopoly status of the local exchange network is now being challenged by (1) continuing rapid technological changes and (2) further rethinking about the scope of the monopoly on the part of policy analysts and regulators. Consequently, as a result of actions at the federal level (in terms of interstate services) and by *selected* public utility commissions at the state level (in terms of intrastate services), competition is emerging in the transport portion of the local exchange network and, to a very limited extent, in local switching. Furthermore, from a technological perspective at least, there are even some prospects for intermodal competition (e.g., from wireless technologies alone or from a combination of cable television and wireless technologies) for the local loop itself.

The cable television industry is now deploying fiber in the distribution portions of its networks and otherwise studying and beginning experiments with the next generation of cable technology. Those developments and experiments, which include "fiber to the neighborhood," conversion from analog to digital transmission, digital compression, and advanced forms of packet switching, are aimed at providing dramatic increases in capacity (e.g., to 500 television channels or beyond) and increased reliability while serving as a future platform for the introduction of two-way, multimedia services (such as interactive games and distance learning) and, potentially, traditional voice and data services as well.

However, there are numerous technical uncertainties and other constraints associated with the provision of the latter types of services. For example, there are the noise ingress problems associated with the use of the "tree and branch" cable architecture for the upstream transmission of such signals, the limitations associated with inexpensive cabling and other devices (e.g., splitters) used by many consumers within their homes, the added complexities in terms of the radio frequency modem needed on the customer premises to allow two-way communications and in interfacing existing analog telephone equipment to that modem.

Other major constraints are that cable companies lack (1) the necessary switching systems to route ordinary telephone calls among subscribers or between subscribers and long-distance networks; (2) the sophisticated and very specialized network management (operational support systems) for the provisioning, administration, maintenance, traffic management, service evaluation, and planning and engineering associated with switched services; (3) specialized billing systems capable of handling the full range of local exchange telephone services and the associated large volume of individual calls or transactions; and (4) the technical arrangements for routing traffic between and among individual local cable television systems and between such systems and the incumbent local exchange telephone carrier.[1]

In addition to the above limitations, the following points must also be considered when considering competition by the cable industry:

- The industry is by no means unanimously in favor of pursuing even specialized aspects of the telecommunications business, let alone ordinary local exchange and exchange access services. In other words, some cable firms may be very slow to offer (if ever) anything except entertainment video and closely related services.

• Because cable television franchises are typically awarded on a community-by-community basis, the industry is badly fragmented on a geographical basis. In a typical metropolitan area (let alone over an entire local access and transport area served by a Bell operating company), it is not unusual to find literally scores of individually franchised systems owned by a relatively large number of different cable multiple system owners.

• Those firms that are exploring the telecommunications business (especially switched services) have not reached any final agreement on the technical details of the architecture that they will employ. Because of the differences in approaches in this early stage of development and the diversity of cable system ownership in a particular region, the initial services offered are apt to be highly fragmented and therefore of limited utility.

• While the economics for upgrading cable networks to expand entertainment services are well established, there is still considerable uncertainty about the economics of providing ordinary switched telephone services.

• Building on the previous four points, the well-publicized trials that the cable companies are currently carrying out are more in the nature of experiments as they relate to the provision of ordinary telephone (i.e., nonentertainment) services.[2] To the best of my knowledge, there are no cable television companies in the United States that are offering regular "dial-tone" services on a tariffed basis at this time [Fall 1993].[3]

In addition to the above, and beyond any direct de jure barriers to entry, there are a host of other de facto constraints to cable companies and other potential competitors entering the local exchange telephone business. These restrictions include, inter alia, the following:

• Lack of agreements and standards for the efficient and economical interconnection of competitors with the incumbents' local exchange networks for the exchange of *local* traffic;

• Difficulties associated with accessing pole lines, ducts, rights-of-way, radio sites, and buildings necessary to construct competing facilities;

• Absence of nondiscriminatory access to unbundled information (e.g., data bases) and services in the incumbent carrier's network necessary for the transmission and routing of telecommunications services;

• Absence of the ability to resell the incumbent's telecommunications services and network functions and skewed pricing structures that preclude entry into some segments of the local exchange market; and

• Incumbent's control of numbering plans and lack of local number portability.[4]

Because of these restrictions and the simple size of the task required to duplicate the existing local exchange network, the cable companies are not in a position to offer significant competition to incumbent local exchange telephone companies in the short to medium term, even if such competition were authorized by regulatory authorities.

Telephone Company Competition with Cable Companies

In the past, local telephone companies or local exchange carriers (LECs) have been limited technologically in their ability to provide cable television services by the inherent bandwidth limitations of the twisted-pair copper cable deployed in the local loop portion of their networks. The LECs are exploring two different approaches to overcoming these limitations.

One approach is known as asymmetrical digital subscriber line (ADSL) and is based on the use of sophisticated digital signal processing techniques for one-way transmission of broadband digital signals plus ordinary two-way voice conversations over a single pair of wires (i.e., the local loop). The ADSL technology, which would retain the classic star topology of the existing network, coupled with advances in digital signal compression, would, potentially, allow simultaneous transmission of several video signals to the home.

Basically, the ADSL technology involves trying to squeeze the last increment of capacity out of a network that was optimized for narrowband voice transmission. However, there are significant technological and economic risks associated with further development and deployment of this technology. These risks are as fundamental as whether the quality of the signal will be sufficient to meet customer needs, whether a sufficient number of video signals can be delivered simultaneously, whether the performance will be adequate on the copper pairs installed many years ago, and, since there are distance limitations associated with the technology, how customers beyond a certain distance from the central office will be served. These risks are apparently so great that some telephone companies in the United States appear to have rejected the ADSL approach, and only one company seems to be strongly pursuing the technology in field trials.

The other approach being explored by the LECs to overcome the inherent limitations of their existing network involves replacement of the twisted-pair copper cable with fiber optic cable and its associated electronics. The telephone companies' highly publicized original vision called for running fiber optic cable all the way to homes, the so-called fiber-to-the-home (FTTH) architecture. However, as noted in a recent article in a telephone industry trade publication (Wilson, 1993), "The local exchange carrier community's once-grand plans for running fiber optic cable into every American home came to a screeching halt in 1989, colliding hard with the wall of high costs" (p. 9).

The industry retreated from the FTTH vision to a configuration that involved proposing to run fiber close to a group of customers so that the cost of the necessary electronics could be shared among a number of households. This configuration, which is now being tested, is referred to as fiber to the curb (FTTC). However, there are still significant unresolved technical issues relating to this more modest vision of the telephone companies. These include such fundamental questions as whether the network (1) should be active or passive or some hybrid of the two, (2) should utilize analog or digital transmission or some hybrid of the two, (3) should be configured with a star or tree-and-branch topology or a logical hybrid of the two, and, (4) for the star topology, whether it should be a single or double star.

This uncertainty is further illustrated by recent indications that some LECs may retreat even further and adopt a "fiber-to-the-neighborhood" architecture that in many ways is similar to the hybrid approach being deployed by the cable industry. Given the technical uncertainties associated with FTTC, it is not surprising that there are considerable economic uncertainties associated with the FTTC approach, to say nothing of FTTH.

Not only are there still significant uncertainties associated with the technology and costs associated with potential telephone entry into the provision of broadband services, there are also significant uncertainties about whether the added revenues generated by the provision of entertainment video services would be sufficient to cover the added costs. For example, one of the most publicized telephone company experiments with two-way cable (GTE's experiment in Cerritos, California) was recently characterized as a "flop" (Lippman, 1993).

Moreover, in competing for the delivery of entertainment video programming, the telephone companies face cable companies that have already constructed broadband coaxial cable networks that are now available to roughly 96 percent of the television households in the United States. Furthermore, as suggested above, cable companies can greatly expand their available capacity and achieve significant one-time efficiency gains by introducing fiber in the trunk and distribution portions of their networks, while retaining use of the existing broadband coaxial cable connections to individual residences. Finally, the cable television revenues themselves are at risk because of the increased competition from terrestrial-based multichannel multipoint distribution systems or "wireless cable" systems, the imminent launch of a high-power, high-capacity direct broadcast satellite system, and the "cellular-like" wireless systems operating in the upper regions of the microwave spectrum near 28 GHz. These competitive threats are enhanced by the availability of the same digital compression techniques that will enable the significant expansion of cable television system capacity as explained above.

The issue of LEC investment in broadband facilities is complicated by the interaction of the incentives created by the mature state of the market for ordinary local exchange services, continuing rapid technological changes, and the nature of traditional economic regulation. Regarding market maturity, one of the most fundamental factors facing the LECs is that "universal service" has largely been accomplished and their basic business, as measured in terms of the number of access lines, is growing at the slow rate of only a few percent per year. Thus, there is very little opportunity for substantial growth in their basic business.

Moreover, their basic local exchange telephone business is, by and large, still under traditional rate-of-return regulation. With traditional rate-of-return regulation, expenses are essentially passed through to ratepayers, and annual profits are determined by the product of the allowed annual percentage rate-of-return and the net investment (i.e., original investment less accumulated depreciation) that has been made to provide the regulated services. However, the falling costs of switching and of interoffice transmission, and the introduction of pair-gain systems in the loop, combined with more rapid depreciation, have created a situation wherein the net investment (i.e., the "rate base") is trending downward. In a nonregulated industry, this would be welcome news because increased capital productivity would allow increased profits and/or decreased prices. But with traditional rate-of-return regulation, increased capital productivity reduces the absolute level of profits. This decrease in the absolute level of profits due to falling net investment has been exacerbated by the falling cost of capital (and hence the allowed rate of return) over the last few years.[5] In short, without some fundamental change in strategy or regulation, the LECs/BOCs face the prospect of "disinvestment" and *decreasing* total profits in their basic business.

Thus, as long as the established LECs are rate-of-return regulated—or even if a strong component of rate-of-return regulation remains under alternative forms of regulation as is typically the case—there is a strong reason for them to invest in fiber irrespective of demand. Indeed, the potential drop in the absolute level of profits without the immense investment associated with fiber-in-the-loop may largely explain the massive public relations, litigation, and lobbying efforts that they have unleashed to convince policymakers, regulators, and the general public that the investment is not only prudent but absolutely required to assure the competitiveness of U.S. firms in world markets.

There are other advantages to the BOCs of deploying fiber, especially in downtown metropolitan areas where they are beginning to face competition from competitive access providers such as Teleport and Metropolitan Fiber Systems. By investing in fiber in these downtown areas, they not only respond to the nascent competition but, by over-investing, can also discourage additional entry. Despite these other advantages, the need to boost investment is clearly a major motivation for the LECs' desire for fiber—especially in the residential and small-business markets.

Another problem with LEC investment in fiber is that, as long as they have monopoly power in the provision of local exchange services, they have the incentive to load the bulk of those investment costs on ordinary ratepayers in order to subsidize the provision of new broadband services. In addition, as a practical matter, it is almost totally beyond comprehension that regulators would allow a local telephone company to fail if there were not alternative providers of service available. Thus if uneconomic investments by LECs are made in broadband facilities in the subscriber access portion of the network, the inevitable result will be that ordinary ratepayers—not the stockholders of the local exchange company—will be stuck with the bill. This is especially true if such investments are made at the behest of the regulators themselves.

Because of all of the foregoing and the simple size of rebuilding the existing local exchange network, the telephone companies are not particularly well situated to compete with the cable companies for provision of broadband entertainment services *unless potentially uneconomic investment is artificially stimulated by policy and regulatory actions (or inaction)*.

SUMMARY AND CONCLUSIONS

Despite the current hype that surrounds the issue, there are significant uncertainties about whether head-to-head competition will actually develop between telephone companies and cable companies in the provision of local services. Even if the existing legal barriers to entry are eliminated, there remain significant technological, economic, and marketplace uncertainties concerning the development of such competition. If competition does develop between the two industries, it will be slow in coming, and it may well be limited to the provision of certain specialized services or network components, rather than for the basic services that lie at the heart of each industry's current business. Furthermore, if, over time, these barriers and uncertainties are overcome, the resulting industry structure may consist of little more than an oligopoly with only two major players.

In either case—with limited competition in adjacent sectors of the market or a two-player oligopoly—the resulting rivalry would hardly meet the test for robust competition that would justify full deregulation of the LECs. Thus, policymakers and regulators would be well advised to exhibit a healthy amount of skepticism regarding the lifting of existing line of business restrictions and deregulating LECs without adequate safeguards.

REFERENCES

Lippman, John. 1993. "'TV of Tomorrow' Is a Flop Today," *Washington Post,* September 1, p. F1.

Wilson, Carol (ed., Telefocus column). 1993. "Bellcore Revisits the Residential Broadband Cost Question," *Telephony,* July 26.

NOTES

1. Affadavit of Richard R. Green in the United States District Court for the Eastern District of Virginia, *The Chesapeake and Potomac Telephone Company et al. v. United States of America et al.,* May 20, 1993.
2. Affidavit of Richard R. Green, May 20, 1993.
3. In the United Kingdom, cable companies are beginning to provide ordinary local telephone service in competition with British Telecommunications.
4. These barriers were identified and discussed at the Eighth Annual Aspen Institute Conference on Telecommunications Regulation.
5. The potential decrease in profits also explains, to a substantial degree, the LECs/BOCs interest in "price cap" or "incentive" regulation, which essentially decouples prices from costs.

Regulation and Optimal Technological Change: Not Whether, But How

Nina W. Cornell

The topic of this panel is how regulation has impeded the timely introduction of cost-effective new technologies and even encouraged adoption of new technologies that, although new, are not the most cost-effective. I have been asked to address this question from the vantage point of an interexchange carrier. However, I believe that the vantage point is the same as that of any entrant that wants to provide services but is not the provider that supplies the network with which end users initially connect. The problem I have with the formulation of the panel's topic is that it implies that if regulation somehow went away, a more efficient future—with speedier and better technological change—would await us. I do not believe that.

The technological change that an industry or sector experiences is in part a function of its market structure and the presence or absence of barriers to entry. In telecommunications the choice for market structure is either to have a monopoly or to use regulation to keep the door open to the possibility for multiple-firm supply. Which of these two choices we select and, if we select the second, how we implement it will have profound effects on the future pace and direction of technological changes.

I believe the history of telecommunications from its beginnings to the end of the monopoly era shows that a monopoly market structure does not maximize the likelihood of promoting the most cost-effective technology at the most efficient pace. As the telecommunications system was opened up to entry in different places, starting in the 1970s particularly, the pace of deployment of new technologies and functionalities also accelerated.[1] Only if we maintain and even expand the opportunities for multiple-firm supply and new entry will the goal of promoting the most efficient pace and direction of technological change be met. To maintain or expand the opportunities for multiple-firm supply, however, requires regulation in its broadest sense. This is because telecommunications is a two-way network system.

In all two-way network systems, government has had to play a bigger or smaller role in order to ensure interconnection between the parts of the system. This is because two-way networks must have some kinds of nodes where the different parts of the network pass traffic in both directions between the parts. If any one player completely controls the nodes, that player also controls the entire system. Any would-be entrant would have to build a complete duplicating network and induce all customers to switch, as otherwise the new network would not serve the two-way function that is required.

In order to have the possibility for multiple-firm supply when the sector involves a two-way network need, government must either own the nodes, as it does with airports; own the approaches to nodes, as it does with roads; or regulate the terms and conditions of interconnection, as it does with railroads and telecommunications. If government does not continue to regulate the terms and conditions of interconnection in telecommunications, the history of this sector of the economy shows that those terms and conditions will be used to remonopolize most or all of the sector.

Technically, this outcome is possible because of the way the two-way distribution of traffic occurs. Each end-user connection to the network (or to a network of networks) must have a unique address number, whether that number is the one originally dialed (as with current ordinary local and toll calls) or is the number to which a dialed number is translated for routing (as occurs today with 800 calls). This unique address number is what the networks themselves use to route calls. When a call is placed by a calling party, that unique address number is set to be the termination point. Most customers, particularly residential and small-business customers, will have only one or a very small number of such terminating addresses and are likely to want them all supplied by a single company, even should local exchange competition be a reality. I base this conclusion on the way the toll market has worked: almost all interexchange companies use the billing services of the local exchange companies for residential customers, at least in part because of those customers' preference for a "single bill." Moreover, at least for a while, even where a customer uses more than one local exchange supplier, the telephone number actually dialed by the calling party will determine which of those lines is the destination point.

The company that controls that termination point has dominance over the pricing of terminating at that address within a very broad range of possible prices. For most kinds of what we today call plain old telephone service calls—local and toll calls—the company that receives the origination of the call will have no choice but to pay what the company that controls the terminating address charges for the termination. This is because the choice of which company will control the terminating address is not made by the company of the customer originating the call but by the end-user customer on the basis of the cost he or she has to pay, which is only for outgoing calls. The cost to the originating company of trying to convince the terminating customer to change that choice is not likely to be low enough that the originating company could afford to try to induce the terminating customer to change his or her provider of that connection to the network. Thus, the company that controls the terminating address will continue to have a bottleneck monopoly over all calls sent to that address, even if not over all customers in a geographic region.

Regulation of the terms and conditions of interconnection will have to take into account a number of ways the company that controls the terminating address might use (abuse) that control, absent regulation, to remonopolize the market, if multiple-firm supply is to be able to bring new technologies to market tests. The first way the company might try to regain a monopoly is to refuse outright to interconnect at all. This is also the easiest abuse to prevent. As important for future possible technologies, however, is the potential refusal of the terminating company to interconnect except at a few select locations in its network. This selectivity can be used to retain a monopoly over a larger amount of the actual transmission or to insist that additional functions will only be supplied by the terminating company, not the originating company. A refusal to interconnect at a particular point in the network can prevent certain technologies from being introduced at all or from being introduced in as timely or as extensive a manner as would have occurred if interconnection had been allowed where the supplier of that technology had wanted.

Refusals to interconnect are not the only way power over the nodes can be exercised to influence market structure and, with it, technological outcomes. The same results can be achieved by allowing interconnection but then setting prices that in fact deter or prevent the ability to use the interconnections successfully. In this regard it is not just the setting of too high a price, per se, but also the relationship of the price for interconnection to other prices that can have pervasive effects on the future pace and direction of technological change. The adverse effect of cross-subsidization on future technological change has been touched on by Robert Crandall (in this volume). Crosssubsidies by an incumbent firm with control over the nodes can be used to impede or block entry by other firms into the provision of one or more services. This in turn prevents those other firms from bringing new technologies and new functions to the market.

Much less often addressed, but a much more common danger, is the likelihood of establishing price squeezes that achieve the same goals but without having to price any services to end users below cost. The formerly monopoly local exchange providers have succeeded in several cases in establishing price squeezes and getting to continue them, reducing or eliminating the ability of other firms from effectively challenging the former monopoly provider's dominance over the supply of those services. Price squeezes by local exchange carriers have been demonstrated in the case of pay phones in a number of state regulatory proceedings.[2] The battle that continues to rage over whether local exchange carriers will be required to impute charges for bottleneck monopoly functions into their end-user rates also demonstrates the existence of price squeezes today.[3]

Local exchange carriers—for example, GTE in Oregon and Pacific Bell in California— continue to argue that they should not be required to impute the charges at all.[4] Moreover, they argue that if they must impute at all it should be done over such a broad aggregation of services as to allow some of the most competitive services to escape the requirement. This argument appears, for example, as a claim that any imputation requirement should not apply on a service-by-service basis within intra-local access and transport area (LATA) toll, but only to the entire category of toll services collectively. Because of the continued denial of equal access for intra-LATA toll, the most competitive portion of this market is the segment that interexchange carriers mostly supply using special-access facilities. An imputation requirement applied only to all toll services collectively imposes no effective constraint on how the local exchange carrier prices the services that compete with the interexchange carriers' special access-based services. When imputation is not required, the result is that the local exchange carrier can price the service at a level no other carrier can match because of the price that the other carriers must pay for access.

This is the same technique that will most likely be used to remonopolize if proper regulation over interconnection prices is not maintained even in a world of multiple suppliers. This is particularly important because multiple suppliers are not likely to be able to enter with services offered to all customers in a broad geographic region. The incumbent supplier of network connections—the local loop—is likely to try to impose interconnection rates on competing providers that are higher than what it recovers in its own charges for the service for which termination is sought, as Bell Atlantic tried to do in Maryland, for example, for local exchange terminations to Metropolitan Fiber Systems.[5] If permitted to price this way, the current providers of local loop facilities could underprice all other providers in all markets in which they are allowed to operate.

Thus, the choice facing the government in telecommunications is not whether to regulate but where to regulate, and how, if it wants to promote the ability of the market to test and adopt most readily new cost-effective technologies. Whatever answer government chooses will have a significant influence on the future pace and direction of technological change, because it will determine into which parts of the whole telecommunications system new ideas, technologies, and services can most easily be subjected to a market test. In other words, regulation cannot be neutral in its impact on the future pace and direction of technological change.

Regulation will have to determine the nature and extent of physical unbundling that will be required of those firms that supply the first connections to end users. How unbundled those networks are forced to be—which means at what points other firms can interconnect—will determine how much other firms can contest in the market such functions as the amount and direction of the intelligence in the network, the functions that can be added to basic call processing, and the like.[6] Other firms can only contest the choices made for these elements if they can interconnect at locations that enable them to offer substitutes for the choices made by the initial carrier.

Regulation will also have to determine the prices paid for various forms of interconnection. In this regard, regulation will have to determine three aspects of the price. The first is the overall level of each price. The second is the price for one form of interconnection relative to other forms of interconnection. The third is the price for each form of interconnection relative to prices set by the owner of the nodes for services provided directly to end users. Each of these aspects of the price for interconnection can have a direct impact on the future pace and direction of technological change.

If interconnection prices are set very high, demand for the services that must interconnect is suppressed, with the result that less attention may be paid to those services by those investigating new technologies and functions. Very high interconnection prices could also push some users to turn to less efficient technologies that can serve their needs but that avoid the need to pay the high interconnection prices. Because such options are only available to some telecommunications users, the result is a loss in efficiency over time.

If the price for one form of interconnection is high relative to another, firms using new technologies will tend to try to adapt those technologies to be able to use the less costly (to them) form of interconnection. This also will have direct impacts on which technologies get adopted and how efficiently they are used. If the relative prices for different forms of interconnection vary based on factors other than the underlying cost differences, these variations in how future technologies develop will not be as efficient as they would have been in the absence of interconnection price differences not based on cost differences.

If the price for interconnection is set such that any firm using interconnections faces a price squeeze relative to the price charged for relevant end-user services by the firm controlling the nodes, the firm using the interconnections will have less chance to prove in the market the possible superiority of its technology or functions. Once again, the result will be less efficient than would have been the case in the absence of a price squeeze and will have a deleterious effect on the pace and direction of future technological change.

The challenge, thus, is to accept that some regulation is necessary and to design it to allow the greatest opportunity for new ideas to have a chance to prove themselves in the market. The necessary regulations involve both the technical aspects of interconnection and the prices to be paid for various forms of interconnection. The first of these two needs means that regulation must address the multitude of issues involved in what has come to be called Open Network Architecture, but it must be done with a greater recognition than appears at present of a need to push the former monopoly firms that provide the first connections to the system for end users to unbundle beyond where their own incentives would lead them to go. The necessary regulation of interconnection prices, however, will not occur until there is recognition that the necessary controls over interconnection prices are not addressed by debating the merits of price caps over rate-of-return regulation.

The problems of rate-of-return regulation are by now widely known. Price cap regulation is offered as a solution that eliminates incentives to avoid new technology or be inefficient in operation. Price caps are also advocated because they may reduce or eliminate incentives to engage in cross-subsidization. This latter claim depends in good measure on the ability to prevent the firm from turning back to rate-of-return regulation if things do not go well for it under price caps, a topic most suitable for debate among lawyers. Price caps, however, do not eliminate the incentive to create a price squeeze and may even facilitate implementing one. Thus, price caps, however beneficial as a means of limiting possible monopoly abuses of end users' prices, do not begin to address the issues involved with pricing interconnections. Moreover, the regulatory requirements necessary to try to control or prevent price squeezes reduce some of the regulatory savings that have been touted as a benefit by those who promote price caps.

The controls necessary to try to curb or prevent price squeezes involve looking in fairly detailed ways at costs versus prices for each service offering, particularly the offerings to end users. The reason is that, to prevent price squeezes, end-user prices must include the same charge for use of the nodes as is charged to those who must interconnect. Without such a nondiscrimination requirement, price squeezes are inevitable. The necessary data from which this part of the necessary regulatory controls must be developed, however, is either the same as, or similar to, the data that have to be collected for rate-of-return regulation. Ending the collection and reporting of these data is one of the "efficiencies" often cited as a benefit of price cap regulation. It is, however, at best only a short-run efficiency if it means that the regulations necessary to promote the widest testing of new ideas and functions cannot be imposed.

NOTES

1. See, for example, *Report by the Federal Communications Commission on Domestic Telecommunications Policies,* submitted to the Subcommittee on Communications, Committee on Interstate and Foreign Commerce, U.S. House of Representatives, September 27, 1976; and *Telecommunications in Transition: The Status of Competition in the Telecommunications Industry,* a report by the Majority Staff of the Subcommittee on Telecommunications, Consumer Protection, and Finance, U.S. Congress, House, Committee on Energy and Commerce, 97th Congress, 1st session, November 3, 1981, for two reports of these effects compiled at the time the system was being opened up to entry.

2. See, for example, Before the Commonwealth of Massachusetts, Department of Public Utilities, Petition for an Advisory Ruling as to the Competitive Nature of Public Pay Telephone Service, D.P.U. 88-45, 1988-1989; Before the Public Service Commission, State of Florida, In re: Petition for Review of Rates and Charges Paid by PATS Providers to LECs, Docket No. 860723-TP, 1990; Before the State of Illinois, Illinois Commerce Commission, In the Matter of Independent Coin Payphone Association and Total Communication Services, Inc. Complaint to Reclassify Illinois Bell Telephone Company Pay Telephone Service as a Competitive Service in Illinois Market Service Area 1 (MSA 1), Docket No. 88-0412, 1991-1992; Before the Washington Utilities and Transportation Commission, *Northwest Payphone Association et al. v. US WEST Communications, Inc.*, Docket UT-920174, 1993.

3. See, for example, Before the State of Illinois, Illinois Commerce Commission, In the Matter of Illinois Bell Telephone Company Petition to Regulate Rates and Charges of Non-Competitive Services Under an Alternative Form of Regulation, Docket No. 92-0448, 1993. This is only one of many cases where a price squeeze has been shown to exist for some toll services.

4. See the filings of GTE in Before the Public Utility Commission of Oregon, In the Matter of the Investigation into the Cost of Providing Telecommunications Services, Phase II: Unbundling and Pricing Issues, Docket UM 351, ongoing, and the testimony and briefs of Pacific Bell in Before the Public Utilities Commission of California, In the Matter of Alternative Regulatory Frameworks for Local Exchange Carriers and Related Matters, I.87-11-033, 1991-1992.

5. See the Direct Testimony of Alfred E. Kahn and William E. Taylor on behalf of Bell Atlantic-Maryland in Case No. 8584, Before the Public Service Commission of Maryland, In the Matter of the Application of MFS Intelenet of Maryland, 1994.

6. For a vivid example of the difference in the level of unbundling sought by entrants and the level local exchange carriers are willing to provide, see the filings of the various parties in Before the Public Utility Commission of Oregon, In the Matter of the Investigation into the Cost of Providing Telecommunications Services, Phase II: Unbundling and Pricing Issues, Docket UM 351, ongoing. This is not the only example of such a discrepancy, however.

The Future of Telecommunications Regulation: The Hard Work Is Just Beginning

Thomas J. Long

INTRODUCTION

This is a time of dramatic change in the telecommunications industry. Traditional monopoly markets are disappearing. Competition in the local loop is coming. Computer and telecommunications technologies are converging. The cable industry is becoming a major force in telecommunications. Each day our newspapers herald another important change.

As we make the transition to a new telecommunications marketplace, particularly as we anticipate enhanced competition, there may be a tendency to assume that it is time for regulators to pack their bags and go on an extended, or even permanent, vacation. My message is that regulators should put the vacation brochures back in their drawers and hunker down for at least 5 to 10 years of more hard work. And, despite all of the changes, most of the hard work will serve the same primary purpose that regulation has always served—protecting consumers and competitors from monopoly power.

A number of trends are now converging to create a serious threat for residential and small-business consumers. The threat is of saddling these consumers with unjustifiably high rates for basic service in a time of dramatically declining costs. Monopoly local exchange carriers (LECs) can use these inflated basic service rates to cross-subsidize their expansion into the video and information services markets. LECs will be able to use the revenue from these excessive rates to gain an advantage over their current and future competitors. Consumers will suffer, not just because they will be paying excessive rates but also because the development of competition will be slowed or even stymied.

At least five trends are contributing to this threat. First, skittish LECs worried about competition are looking to collect as much money as possible from captive customers for as long as possible. Second, regulators may be aiming their sights too far down the competitive road and failing to see the considerable monopoly power that LECs still wield. Third, in a time of tremendous upheaval in the telecommunications marketplace, the LECs and other telecommunications giants can take advantage of their massive resources to argue for a regulatory "paradigm shift" that furthers their interests at the expense of small customers. Fourth, telecommunications giants have shown their adeptness at parlaying regional and national concerns about economic weakness and global competitiveness into much friendlier regulation. Fifth, in the absence of any new tax revenue, monopoly telephone rates are seen as the next best thing to taxes. Unlike tax money, which is used for a public purpose, revenue from higher rates is viewed by the monopoly telephone companies as their money to use as they see fit.

These trends mean that regulators will have to be more vigilant than ever in the coming years to protect consumers from unjust and unreasonable rates. In this paper I highlight two issues

153

that I believe will be particularly important to small consumers. But first, since I have argued that the main purposes of regulation will not change much in the coming years, I want to share my view of those main purposes.

PURPOSES OF REGULATION

I see three traditional purposes of telecommunications regulation. The first, and core, purpose is to protect consumers and competitors against the undesirable consequences of market power. With market power comes the ability to engage in price gouging of customers, to cross-subsidize competitive services through supracompetitive rates for monopoly services, and to engage in other unfair conduct designed to thwart competitors. Regulation is needed to curb such exercises of market power.

The second purpose is to protect consumers and competitors against fraud (and other abuses of customers), unfair competition, and other undesirable behavior that even telecommunications firms without market power might exhibit. Market power can facilitate some of this untoward behavior, but competitive firms can commit fraud and conspire to defeat competitors too. (Monopolies don't have a monopoly on bad behavior.)

The third purpose is to promote universal service. This is a goal of regulation, not primarily to satisfy a collective charitable impulse toward less advantaged segments of society, but rather to reap the enhanced benefits of a telecommunications network to which all are connected. I define universal service as universal access to the level of telecommunications service that is essential for effective participation in society.[1]

As competition increases, these will still be the main purposes of regulation, at least for the foreseeable future. As for the first purpose, the LECs will continue to have market power in many of their market segments for several years, especially for the services used primarily by residential and small business customers.[2] With respect to the second purpose, the challenge of protecting consumers against fraud and other abuses will only increase as the number of competitors and customer confusion multiply. As for the universal service objective, even in a fully competitive marketplace, everyone will be better off if we ensure that essential telephone service is affordable to all groups in society. That will mean continuing to have the general body of customers pay into Lifeline-type funds to support subsidized service for customers needing assistance.

I would like to focus on two issues that are now and will continue to be particularly important to consumers, and therefore should be important to regulators, in the coming years. The first is protecting consumers from becoming the guarantors of the LECs' success in competition. The second is ensuring that consumers don't become involuntary and uncompensated investors in the LECs.

INFLATED MONOPOLY RATES AS A CUSHION AGAINST COMPETITION

As traditional monopoly sectors shrink, the incentives for LECs to flex their monopoly muscle with their local service customers will be greater than ever. The LECs are facing the scary prospect—well known to firms in the competitive sector of the economy—that their financial success may not be guaranteed. Captive local service customers are their only remaining hope of retaining the financial security to which they have become accustomed.

The linchpin of the LECs' efforts to shelter themselves from competition is the following familiar assertion: basic exchange service is priced way below cost and subsidized by the services that are now being opened to competition. Those subsidies may have served valuable purposes in the past, the argument goes, but they cannot be sustained if the LECs are to have a fair opportunity to compete in the sectors thrown open to competition.

The beauty of this argument from the LECs' perspective is that it completely exonerates *them* from any responsibility (such as the possibility of inflated costs) for rates that are not competitive. And it enables the LECs to argue that rate reductions for the newly competitive services should be offset by rate increases to the supposedly below-cost basic exchange services.

This argument that basic services are receiving a large subsidy is being used to justify billions of dollars of monopoly rate increases, a large portion of which have already occurred. Since such large sums of money are at stake, it is more vital than ever that regulators closely examine the validity of the subsidy claim.

Some may be surprised at my suggestion that there is a dispute about whether basic exchange service is receiving a subsidy. I chalk that up to the telephone companies' masterful domination of the public discourse, certainly not to any factual demonstration that basic service is being subsidized. Solely through decades of constant, unflagging repetition in every forum and at every opportunity, the telephone companies have managed to transform an unproven assertion into an unassailable fact.

TURN had the opportunity to examine this "fact" in California's intra-local access and transport area competition and "rate rebalancing" case involving our two largest LECs, Pacific Bell (PacBell) and GTE of California (GTEC).[3] The only way PacBell and GTEC could support their claim that basic residential service was subsidized was by allocating all of the fixed costs of the local exchange network to basic service. Of the total reported costs of basic service of about $25 per month, those fixed costs made up about 80 percent.

The problem with allocating all of these costs to local service is that the local exchange network is designed for and used by all telephone services—including toll and enhanced services—not just local service. The fixed network costs are joint and common costs incurred to provide a variety of services. Assigning all of those costs to basic service is about as fair and economically efficient as assigning all of the companies' executive salaries solely to basic service.

What we discovered in our recent case was that PacBell's $8.35 rate for basic service more than covers the *directly assignable* costs of basic service. The difficult question for regulators is how to allocate the huge pot of joint and common costs that are not clearly attributable to just one service. There is no obvious right way to do it.[4] But when phone companies report as *fact* that basic residential service costs $25 a month (a figure that includes all the fixed network costs), they are at best being misleading. In reality, their cost numbers rely on a lot of highly controversial assumptions, arguments, and economic theories.[5]

Regulators have to grapple with this issue if they want to protect monopoly customers from becoming the guarantors of the LECs' success in competition. They need to be skeptical. Many California regulators did not even know there was any controversy about the cost of basic service; they had been so thoroughly indoctrinated by the phone companies' party line. Regulators need to get their hands dirty by requiring and critically examining studies of the cost of major services and then deciding the fair amount of costs to recover from basic exchange services and how much of the fixed network costs should be allocated to other services.

Once the fair amount of costs to recover from basic service is determined, rates should be set to recover those costs, no more, no less. Once it is clear that basic service is no longer receiving a subsidy, the LECs will not be able to argue for increased basic rates every time a former monopoly service is opened to competition. The idea is to build a wall around basic exchange service rates and make it clear that those rates are not available to fund competition.

RATEPAYERS AS INVOLUNTARY INVESTORS

The second major challenge for regulators I want to discuss is that of resisting the pressure from LECs to make ratepayers involuntary investors in the advanced telecommunications networks that the LECs are now building. In California we are hearing a steady refrain from the LECs that they need rate relief to accelerate the development of the telecommunications infrastructure.

Raising rates (or, just as bad, not decreasing rates as much as would otherwise be justified) to pay for infrastructure is doubly unfair to ratepayers. First, it makes ratepayers fund and bear the risks of investment without offering any prospect of a return on the investment. In a competitive market, firms do not have the luxury of simply raising their prices to fund investment. They have two choices: use internal capital or raise new capital in the debt and equity markets. In either case, investors will demand and be entitled to a return on the investment.

Allowing LECs to collect inflated rates from ratepayers to fund infrastructure investment is like giving them free money. They will use that free money to build advanced broadband networks, which they will claim *they*, not the deep-pocket ratepayers, own. By virtue of their "ownership" of this new network plant, the LECs will argue that ratepayers are not entitled to any of the returns on the investment. If this argument succeeds, the LECs will have shifted all of the risk of the investment to ratepayers while denying them the return that investors in our market economy are entitled to. Ratepayers will put up some or all of the capital for the broadband networks and then be required to pay the LECs for the privilege of using those networks.[6]

Compounding that unfairness is the second problem: if rate increases to fund investment are allowed, the customers who will pay the excessive rates are for the most part not the customers who will be using and benefiting from the infrastructure they will pay for. The purpose of the new broadband investment is not to enhance basic voice service but to allow sophisticated new video and high-speed data services to come to market. (One obvious purpose is to make competitive forays into the cable television market.) Using inflated rates for basic voice-grade service to pay for advanced video and data transmission services is a classic and indefensible cross-subsidy.

The enhanced competition that has been a central goal of telecommunications policy for the last decade could be a casualty of such de facto ratepayer funding of broadband networks. Telecommunications firms that are operating in competitive markets do not have the luxury of charging inflated rates for monopoly service to finance their new investment. Lacking the unfair advantages of the LECs, they may be hamstrung in their efforts to mount a significant competitive threat.

If funding of infrastructure investment is left to the market, will that mean that certain groups of people or certain geographic pockets will not be served by new investment? The argument is made that, for these left-out sectors, ratepayer funding will be necessary to provide the advanced infrastructure.

Regulators need to examine these claims closely and skeptically. With free money available, the LECs have a strong incentive to claim that infrastructure investment is uneconomic in a given sector and will not happen without ratepayer funding. Before monopoly rates are used to fund any investment, the LECs should be required to prove that there is no possibility of using private capital for the investment. If the LEC can make this showing and the regulators authorize inflated rates to pay for the investment, then ratepayers should be treated the same as any other investors and receive their fair share of the returns.[7] I suspect that if regulators made it clear that ratepayers would be entitled to returns on investments they are required to fund, many supposedly uneconomic investments would suddenly become economic.

Before raising rates to foster accelerated investment, regulators should do everything possible to prevent basic monopoly services from cross-subsidizing advanced competitive services. The fairest way for the LECs to recover their costs of providing advanced service is through the rates they charge for those advanced services.

There may come a time when a service we now consider an advanced information age service becomes necessary for effective participation in society.[8] When that happens, under my definition, that service would be considered part of universal service. Groups that could not afford the full cost of the service would be entitled to discounts to make it affordable. The discounts should be broadly funded by a surcharge on all telecommunications services.

However, regulators should be skeptical of claims that we need a dramatic redefinition of universal service now to include emerging broadband services. The LECs and many manufacturers have a strong financial interest in a world of universal broadband, even if most of the population does not need it. Proclaiming access to sophisticated broadband-type services to be part of universal service ahead of any demonstrated strong need or demand would enhance revenues for many manufacturers and for telecommunications carriers, which would benefit from major new markets.

Lofty claims are being made that broadband services will revolutionize education, health care, and citizens' ability to participate in government. There are powerful interests that have a strong financial incentive to make these claims. Decades ago we were also told that television would be a powerful educational force and a vehicle for the expression of a broad diversity of ideas and voices. Television has turned out to be overwhelmingly an entertainment medium that offers precious little opportunity for underrepresented segments of society to get their voices heard. It is possible that, like television, broadband technologies will prove to be mostly a means for enhancing entertainment opportunities.

A more prudent approach to expanding universal service would be to wait to see which new services become essential and use Lifeline-type discounts as needed to make those services affordable for those who need financial assistance.

CONCLUSION

Some of the hard work that lies ahead for regulators could be obviated if LECs adopted a different approach. The hard work I have identified falls mainly into the category of protecting consumers against monopoly pricing. Many LECs claim to welcome fair and open competition. Unfortunately, in the regulatory trenches, they are demanding terms for the onset of competition that use their remaining monopoly customers as a crutch and that hamstring their competitors. I challenge the LECs to live up to their rhetoric and to resist the urge to continue to rely on rate-payers as a deep pocket funding source for their private investments.

When it comes to infrastructure development, consumers would be best served by regulation that promotes diverse choices of telecommunications companies and services. Managing infrastructure improvements should not be the endeavor that keeps regulators busy in the next decade.

ACKNOWLEDGMENT

I am indebted to TURN's telecommunications analyst, Regina Costa, for many of the ideas in this paper. Of course, I am responsible for any deficiencies in the analysis.

REFERENCE

Melody, William H. 1983. "Diversification, Deregulation, and Increased Uncertainty in the Public Utility Industries." *Proceedings of the Institute of Public Utilities, Thirteenth Annual Conference*, MSU Public Utilities Papers. Institute of Public Utilities, East Lansing, Mich.

NOTES

1. My wording of the universal service objective reflects my views about how the "service" part of universal service should be defined. It is not enough just to promote access to dial-tone service. Today, the level of service that is essential to effective participation in society would include affordable calling (to businesses, schools, health care providers, etc.) within one's community of interest. The definition of universal service is fluid. It should evolve as our requirements of and expectations from telecommunications service change. Touch-tone service is a good example. A few years ago touch-tone service was convenient but not necessary. Now many businesses, government offices, and health care providers require touch-tone to be able to take full advantage of their services. As a result, touch-tone should now be considered part of the service that should be universal.

2. The large portion of residential customers who use telecommunications services sparingly will not be an attractive target for competition for many years to come, if ever.

3. California Public Utilities Commission Investigation (I.) 87-11-033, Phase III, the Implementation Rate Design (IRD) phase. As of May 1994, the California commission had not issued its final decision in this proceeding.

4. The method I have seen that would best replicate cost causation would make use of stand-alone cost studies of the various services that use the local loop. The joint and common costs of the local loop would be allocated among these services based on the extent to which each service causes local loop costs to be incurred, as shown by the stand-alone cost studies.

Here is a simple example. Suppose that the cost study for a stand-alone toll network (i.e., a network that included all the elements necessary for toll calling and only toll calling, such as the local loop and switching facilities as they would be designed solely to provide toll service) showed that the local loop portion of such a network would cost an LEC 150 units. Suppose that the local loop in a stand-alone local exchange network would cost 100 units. This would mean that the toll network causes 50 percent more of the local loop costs than local exchange service. In this case the allocation of the joint and common local loop costs should ensure that toll service is assigned 50 percent more of those costs than basic exchange service.

5. One of these theories is that nontraffic-sensitive costs (such as the costs of the local loop) should not be recovered through usage-based rates. This theory is based on a narrow view of economic efficiency. It ignores the inefficiencies and inequities that result when services are not priced to recover the costs they cause. To use an example of current import, assume that a LEC builds a new broadband network that is needed only to carry video services but that incidentally can be used to provide basic voice telephone service. The above-described theory would hold that, if the video services are only priced on a usage basis, the nontraffic-sensitive costs of the broadband network should all be assigned to basic exchange service and recovered in the monthly service fee. The result would be that basic exchange voice telephone rates would be used to recover broadband network costs that voice telephone service did not cause. The prices for voice and video services would be distorted, and many basic voice telephone customers would unfairly be forced to pay the costs of services they do not want. For an excellent discussion of the many issues surrounding allocation of joint and common costs, see Melody (1983).

6. Under this scenario, ratepayers would be equivalent to taxpayers being taxed to build the broadband networks that will be part of the "information superhighway." The U.S. interstate highway system was built with taxpayer money, but at least those highways are publicly owned and taxpayers do not have to pay private owners for using what they have paid for.

7. This would be accomplished by determining the amount of the total new investment that is funded by ratepayers through rates that are higher than they otherwise would be. Based on a required annual accounting of the achieved percentage returns on the new investment, ratepayers would be entitled to that same percentage return on their investment. The ratepayers' returns would be flowed back to them in the form of reduced rates, targeting as nearly as possible the services whose rates had initially been inflated to finance the investment.

8. Determining when a service has become essential for effective participation will not be easy or free of controversy. This will be a sociological judgment, not a judgment that will require the specialized technical expertise that regulatory commissions are supposed to have. Accordingly, decisions about expanding the definition of universal service (which will lead to higher surcharges—i.e., taxes) should certainly involve the legislative process.

The many important issues surrounding the potential expansion of the concept of universal service could certainly be the subject of a separate paper.

Costs and Cross-Subsidies in Telecommunications

Bridger M. Mitchell

One of the fundamental issues in telecommunications regulation is how to devise a method for sharing common costs. Some of the costs of the telecommunications system, including the entire range of services provided over the backbone network, are common to many users and services. The debate about how best to allocate the burden of paying these costs is ubiquitous in telecommunications regulation—and is frequently very confused, especially when the issue is whether a particular service or group of customers is being subsidized by other services and/or users.

As an economic matter, cross-subsidy among users and services is a well-defined concept. A group of customers is being subsidized if their price is so low that the service supplier and its other customers would be better off if the service were discontinued. This circumstance occurs only when the increase in revenues to the telephone company from offering the service is less than the increased costs of providing it.

Several years ago the California Public Utilities Commission collaborated with Pacific Bell, General Telephone of California, and the RAND Corporation to investigate the costs of providing local telephone service using current technologies. The objective was to construct a "building block" model of local telephone service, using data from California, that would estimate the best-practice costs of a local service system, using current technology, for communities of various sizes and other characteristics. In areas where no service was provided, such as a new housing development or industrial park, the model would estimate the total cost of a new local network. In areas in which a backbone network was already in place, the model could be used to estimate the cost of adding more subscribers or more usage by existing subscribers.

Several important findings emerged from this study (Mitchell, 1990), all of which are important in the debate about pricing policy and subsidization within the telecommunications system.

The first important finding was that almost all of the additional costs arising from either more subscribers or additional usage are fixed on a per-subscriber basis. Increases in variable costs arise only from additional usage that occurs at peak periods, which in turn gives rise to a need to increase busy-hour capacity in order to retain service quality. The importance of this finding is that usage-based pricing—such as a charge for each call or each minute of use—would not contribute much to the efficiency of the usage of the telephone system. Indeed, taking into account the additional costs of measuring usage and billing for it, the U.S. system of flat-rate monthly charges for residential local telephone service, including both access and calling, is probably not inefficient.

The second important finding of the California study is that by far the most important factor in causing variation across localities in the cost of service is the distance of the subscriber from the local switching center or, to put it another way, the length of the copper wire connecting the customer to the local switch. The distance of this connection is determined primarily by population density, with customers in small towns and rural areas having, on average, much longer connection distances than customers in large cities.

This finding, too, has considerable policy significance. Local telephone prices are typically based on the average cost of service for all customers within a state or served by a single company with a broad heterogeneous service territory. The practice of rate averaging is extremely important because it discourages entry by local access providers that would deploy less expensive technologies for providing service with long connection distances. For example, recent advances in radio telephone systems, such as the transmission systems used by cellular telephone companies, can provide high-quality local access in many low-density communities at significantly lower costs than the traditional wire-based telephone system.

The third major finding of the California study is that the incremental cost of service is much lower than the average cost of service. That is, adding more subscribers and more usage to an existing system is much less expensive, per additional subscriber, than the average cost per subscriber of constructing a completely new system. Thus, on a statewide basis, the average cost of service in California is about $25, whereas the incremental cost of service in most areas is less than half that amount.

An important implication of this finding is that in most areas local residential service, using present telephone technology, is a natural monopoly. Of course, this conclusion needs to be strongly qualified: it does not apply to central business districts of large cities with numerous businesses that are intensive users of services, to customer groups that seek technical capabilities going substantially beyond the features of the existing network, or to groups that are most efficiently served by radio telephone services, which do not exhibit extensive scale economies. Nevertheless, most residential subscribers do not fall into any of these categories.

All of these findings also shed light on the cross-subsidy issue. They suggest that more intensive users are not being significantly subsidized by less intensive users since nearly all costs are independent of usage. These findings also suggest that most residential customers are not being subsidized, in part because most residential customers do not live in areas with long wire connections to the local switch and in part because the incremental cost of serving one more residential customer in a local area is much lower than the average cost of service. For the most part, subsidies of local service go to residents of high-cost communities with low population densities: rural areas, small towns, and suburban communities with single-unit dwellings on very large lots. And, because in some cases these communities could be served at lower cost by other technologies, even here some of the subsidy is unnecessary.

A common argument for preservation of the existing structure of the telecommunications industry and system of regulation is that it is necessary to preserve universal service because it provides a means for cross-subsidizing users who would otherwise be unable to afford service. The California study casts considerable doubt on this argument. Most users are not cross-subsidized, which means that telephone companies would be worse off if they decided not to serve most customers at current prices. Many customers who are subsidized—notably, farmers, owners of rural second homes, and residents of low-density suburbs—could easily afford unsubsidized service.

REFERENCE

Mitchell, Bridger M. 1990. *Incremental Costs of Telephone Access and Local Use*, R-3909-ICTF. RAND Corporation, Santa Monica, Calif.

Economic Ramifications of the Need for Universal Telecommunications Service

Eli Noam

Among U.S. states, New York has been a leader in promoting competition among firms providing local telephone service. But liberalization does not mean instant deregulation. There is still a messy transition phase to go through, especially on the details of interconnection. The question is, does liberalization of telecommunications mean libertarianism, with no governmental regulatory role? And beyond, if there is free entry into all sectors of telecommunications, what will happen?

It is useful to conceptualize telecommunications networks as shown in Figure 4A. Network size is presented on the horizontal axis. The vertical axis indicates the cost of services provided and their utility to the consumer in terms of dollars. Two curves are shown. One is the average cost curve, and one is the benefit, or utility, curve. The average cost curve implies that a network is, in effect, a cost-sharing arrangement, with the burden of cost collectively assumed by users. Networks have a fairly significant fixed cost. As more users join the network the fixed cost is spread evenly over the various users, and the average cost decreases. But, as the network expands, higher-marginal-cost users join the network and the average cost starts to increase. The result is a U-shaped curve for most cost-functions. The curve for participating in the network increases with the number of people on the network. This is due to the positive externalities of network usage. These externalities tend to flatten out as the network reaches a certain size.

There are a number of distinct regions on Figure 4A. Prior to the point where the two curves intersect, cost is higher than utility. Therefore, somebody—either government or private investors expecting a future return—will have to subsidize the network. Beyond the intersection point, there is a region of self-sustaining growth. Beyond the point where the distance between the two curves is at a maximum, there is a range in which the network, left to the decision process of the users, would not grow, because costs are going up faster than utility and thus the net benefits decrease. Eventually, net benefits become negative when the curves again intersect.

This suggests that there are three areas where there will be problems (Figure 4B). We start with the first region on the left, where there is a critical mass problem. In the early stages of a network arrangement, funds for investment must be garnered because the cost of the system is higher than the benefits returned. In the traditional regime, this investment was done by the network monopolist itself. France Telecom, for example, put ample sums of money into Minitel, even giving away the terminals to its customers. Why? Because eventually the combined critical mass of users would be sufficient for everybody to benefit from the network, including its owner, France Telecom.

But, if that is a reasonable approach for a monopoly, it often will not work in a competitive environment because of the possibility of interconnection. A rival company can interconnect with a company's network and benefit from its rival's accumulated critical mass of customers without having made the up-front investments itself.

161

162

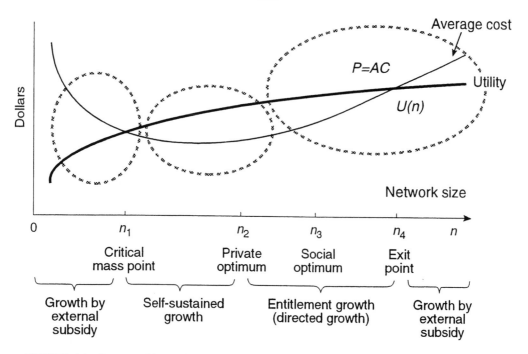

FIGURE 4A Cost and benefits in relation to increasing availability of telecommunications services. *P=AC*, price = average cost; *U(n)*, utility as a function of the number of users.

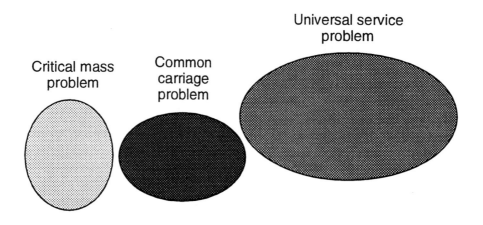

FIGURE 4B Three issues basic to provision of telecommunications services.

Thus, the economic incentives can be in favor of being the second, not the first, entrant. There will be less early investment, and some of the funds necessary to create a critical mass of users might have to come from outside. In Washington and elsewhere, this is an approach technologists are advocating. Many would like to see an injection of government money to deal with the critical mass problem.

But subsidies are not the only alternative. There are essentially three ways to deal with the critical mass problem. One approach for the government is to do nothing and let markets function independently. The second approach is to deal with the cost curve by reducing the initial cost of supplying the service. A final strategy involves demand-side policy, which implies increasing the utility curve by enhancing the value of the fixed services.

The National Science Foundation, by supporting the "industrial policy" approach in the Internet, has both increased demand and reduced cost. Typically, in the United States and elsewhere, the approach has been to concentrate on developing the supply side of the graph (Figure 4A)—for example, by aiding the carriers that would do the upgrading. The alternative, supporting users at the demand side and thus reaching a critical mass, can be done by partly supporting leading-edge, technology-driven uses and also community-based applications. This is where the libraries and schools come in; they do not use leading-edge-technology, supply-side applications but rather demand-side, low-sophistication applications.

The second problem is indicated on the far right of Figure 4A,B, and it concerns universal service. Here the problem is that, because of competition, the network cannot maintain an internal subsidization scheme as in the past with some users subsidizing other users. When entry is free, the new competitors first target the users presenting the largest potential profits, and eventually the support system vanishes. To prevent such a collapse, a very elaborate system of access charges has been established. It is a system that is so opaque that hardly anyone can understand its workings. But in a democratic society there should be accountability, with an understanding about who is paying whom and how much.

There is a relationship between the critical mass and universal service problems. We will never get to the latter without having solved the former. However, it will be harder and will take longer for critical mass to be created unless there is also a willingness to deal with the universal service issue. Consider the Burns-Dole-Gore bill of a few years ago. In essence it said that in order for the government to aid (i.e., to support) by regulation the upgrade of the telecommunications network from copper-wire to fiber (the creation of a critical mass issue), it would also have to ensure that the rural United States would not be left behind (the universal service problem). In other words, it is not realistic to assume that Congress will pour money into the first problem without having some assurances that the universal service side of the network will also be taken care of.

The universal service issue is not going to go away. Some are convinced that competition will take care of any problems that arise. However, they are confused about the difference between *allocational distribution* and *production efficiencies*. Competition will increase production efficiencies. It will not, however, resolve issues of distribution. The fact that food growing and processing are competitive and efficient does not respond to the question of how to feed the poor. Even though it is possible to reduce marginal cost and the absolute number of people not served by market forces acting alone, the universal service issue is not fully addressed that way. There are positive externalities to adding to the network, as depicted by the upward-sloping curve in Figure 4A.

One alternative funding mechanism for universal service is a system that charges all carriers in proportion to their net transmission revenues and credits them for already existing contributions to universal service. The money would go to categories of customers designated by the political system. They would get vouchers for their subsidies and be able to select their own carrier. Competition for such customers would force the carriers to be efficient. Carriers serving low-density areas could also be benefited, although not on an exclusive basis.

A further problem is represented by the middle range of Figure 4A,B. A competitive environment for telecommunications involves a complex arrangement of carriers providing services for different segments of the market: local, mobile, long distance, and international, in addition to vendors of hardware and software. In such a complex carrier system, specialized "systems integrators" will invariably emerge to put together packages of services. This is already happening today for large customers, and will most likely be available to smaller customers in the future. A restructuring plan by the Rochester Telephone Company is going in this direction by separating the service function from the network carrier function.

These systems integrators will effectively undercut the underlying carriers. The basic economic reason is that in a competitive market, the integrators get the services and can resell them at marginal cost, while the carrier has to bear the full cost per unit, including an allocation of fixed costs. Since in network industries with large economies of scale the marginal cost will be below the average cost, the integrator will economically outperform the carrier providing the underlying service. Several things will therefore happen.

One is that the whole notion of common carriage, which is based on similar conditions of service for similarly situated customers, will break down because the common carrier cannot compete against a systems integrator or contract carrier that can price-differentiate and select customers. Secondly, to the extent that the common carrier cannot recover its full cost, there will be under-investment in the network infrastructure itself. In the future, policymakers will have to deal with the question of how the network system can maintain its financing in a situation like this.

In a competitive environment of carriers, systems integrators make unnecessary many of today's regulatory interventions. For example, price regulation becomes largely irrelevant because systems integrators will shop for the best deal for their customers and will compete among themselves for those customers. But what is not going to go away under competition is the universal service issue, and the common carrier (open access) issues. Interconnection problems are also likely to remain in this environment.

The final question is how to reform the existing regulatory system to deal with future issues. Although many people believe that state regulators are part of the problem, it is important to recognize that the decentralized reform process has worked reasonably well in the United States. The Japanese, French, and Germans, all of whom have a centralized decision-making process, have taken decades to reach what a decentralized system has accomplished much faster. Regulatory reform in the United States might not be a logical progression based on a national blueprint, but in most other places these blueprints have taken long to draft and are invariably timid. The United States, using a decentralized mechanism, has been able to reform its telecommunications system quite well. Those who believe that a decentralized Adam Smith-type mechanism is working well in decision making in the sphere of economics should not reject out of hand the same mechanism for decentralized decision making in the political sphere.

In conclusion, liberalization of telecommunications does not mean libertarianism. Many regulatory issues will go away, but others will persist in new forms and new ones will emerge, as discussed. For better or worse, government will not vanish from this sector; the challenge is to orient it toward an open future.

Discussion

BRIAN KAHIN: I want to pick up on Nina Cornell's concerns about interconnection and the issue of the government's role. It is worth bringing up because it is very often overlooked. It is not even mentioned in the administration's information infrastructure document. The NSF [National Science Foundation] is involved in the program to restructure the NSFNET, which is in effect unbundling the NSFNET as it exists today so that NSF is no longer providing a single package in its main contract but has broken out the different roles of routing arbiter, a very high speed backbone that will be very restricted in use, and the newly conceptualized NAP [network access point].

In principle the NAP is an exchange or switch where anybody can interconnect, and it is a way to get onto the very high speed backbone if you have the right to do that. It is not clear how many NAPs there are going to be. NSF is going to fund them, and so we have the government funding something that is not subject to an acceptable use policy but that is general-purpose infrastructure investment.

To what extent will this architecture catalyze the larger infrastructure? How will the commercial Internet crystallize around this concept? I wonder if we are seeing an inversion of the paradigm from the network as a carrier to the switch as a carrier and what that means for understanding the cost-price differential, because it seems to me it is a lot easier to get at that when the carrier is as simple as a switch. That is not going to solve the "last-mile" problem, but it might be progress.

ROBERT CRANDALL: I still have some difficulty with imagining what the network or networks of the future are going to look like. I do not see any reason why there is one network, given what we know about what is happening to transmission switching costs. I can imagine lots of different networks into which I could connect with a piece of terminal equipment that has a set of switches that can connect me to lots of different networks and lots of different nodes.

One of the problems of this discussion is that it is based on what we know is the current network, the current public switched network developed under regulation. One of the things we know about other industries is that once we deregulate, the technology changes and market organizations change in ways we could not possibly predict. In fact, it may be that regulation is condemning us to one particular model that is incredibly inefficient.

KAHIN: Also, the network that we have been focused on is a homogeneous network, whereas now, especially in the Internet, we are entering an environment of heterogeneous networks, where the networks all differ in capability and functionality and speed.

CRANDALL: That is right, but there could be a bunch of different backbones, too, not only a bunch of different LANs [local area networks].

ALFRED AHO: This panel spoke of some of the experiences learned from regulation in Japan. In the United Kingdom, there is ongoing competition between the telephone companies and the cable companies. Are there any lessons that we can learn from regulation in the United Kingdom?

CRANDALL: I do not know much about what has happened with cable versus telco competition, but they have admitted that going from one carrier to two carriers—but making sure there was not a third, fourth, fifth, or sixth carrier—was a mistake, that a duopoly was not much better than a monopoly if it was protected by regulators from entry.

ROGER NOLL: In the United Kingdom the regulatory system itself has not led to competition of the form that we think of in most of the U.S. system. It has led to a somewhat better but nonetheless regulated system whose prices are somewhat better than they were but are not as low as ours. Bob, do you want to talk about that?

ROBERT HARRIS: Yes, just very briefly. The joint offering of telephony and cable entertainment services in the United Kingdom is very different because they basically did not have a cable system. The cable system being deployed was originally designed and engineered to provide both telephony and video and hence does not have a lot of the problems that Dale Hatfield talked about that the U.S. cable system has. It does suggest, however, that as we modernize cable plants in the United States, they can perhaps be reengineered if enough cable operators get the idea that there may be a market for telephonic-type services. But in comparing a de novo situation in Britain to an existing cable plant in the United States, I think you had better be very careful about any kind of lessons you draw from one to the other.

NOLL: I think the main lesson Americans have tried to draw from the United Kingdom concerns how they regulate prices, and there are two lessons to draw on. The first is that they have price regulation; that is to say, all prices taken together, indexed, are capped by the rate of inflation minus 3 percent. It turns out that that system has demonstrated two important principles. The first is that the rate of technological change and advances in productivity inside British Telecom substantially exceeded historical experience, and the second is that the politics of the system requires that they change the rule to make prices fall even faster. That is to say, the second important lesson is that the notion that you can have price regulation that leads to better price performance but enormous profits of the regulated utility is obviously incorrect on the basis of the experience in Britain.

COLIN CROOK: One interesting early indication in the United Kingdom in the cable area is that when the cable companies had basic telephony on the cable system, the turnover on the cable side went down to essentially zero. People will not disconnect their telephones like they will disconnect their cable systems, so some very interesting business is going to be lining up when we start to merge these areas together, I would forecast.

GEORGE GILDER: Two questions. First, it seems to me that digital wireless technology is going to destroy that differential between the cost of rural and urban service almost entirely, and I wonder what Dale Hatfield, for example, thinks about that prospect and what its effect will be.

Second, it seems to me that the price of nodes drops about 50 percent every 18 months and that they just multiply increasingly and that nodes are no longer a bottleneck. Nodes are dropping radically in price and moving rapidly to the fringes of all the networks, so the nodes are really in the command of all the users rather than in the command of the transport people, and I wonder how this process is going to affect the vision of the node as a bottleneck that allows monopoly power to be exercised.

DALE HATFIELD: Let me comment first on wireless. I think that there are some very attractive possibilities with wireless, so I think you are absolutely on the right track. However, you are still faced with most of those other issues I talked about, the terms and conditions of interconnection with the existing carrier and so forth. Also, there are probably some bandwidth issues, perhaps obviously less in rural areas, but to the extent that we are getting more bandwidth-intensive applications, even fairly generous spectrum allocations tend to look not quite so generous, and then I would also say it is not clear to me how all this will sort out.

At one time, I looked on cellular as perhaps having some prospects, but it ended up being in relatively concentrated hands. With the spectrum that the [Federal Communications] Commission has now set aside for PCSs [personal communication systems], it is not clear to me who is going to end up winning that spectrum, and it is not clear to me that we will even then have a fully competitive marketplace. Generally the conclusions that I had before, which I could summarize by saying I am uncertain, still hold: I am still uncertain, even with the prospects of wireless, for those reasons.

NINA CORNELL: I think there is one thing that is not well recognized. Radio may help with the rural-urban distinction, assuming you solve the bandwidth problems and the concentration problems. But to serve a location like the one where I live, you are talking about having to put in a number of radio transmitters to serve essentially 10 households, and so you are still going to have a very high cost per household because there is very little sharing of a particular set of technology. So you are still going to have that same phenomenon of, if you pardon the expression, "dis-economies of scale" for rural, because even though there are a few lovely places, probably in South Dakota and Nebraska, where they can go long hops and not meet mountains and convoluted terrain, a lot of the most sparsely populated areas happen to be in the mountains.

The second thing about nodes is that it is not that the price or cost of nodes is so high; it is literally that they are—despite price, despite cost—a bottleneck because you have to get from the node to the end user. It is that link, if you will, and its connection at the node, that has been used so successfully—and continues to be able to be used so successfully—to control the system if there is not some regulatory control in the broad sense over how it is used and what the prices are. Even if I had continued to live in downtown Washington, I would not have the ability to call up and say, "Please connect my local loop to somebody else's node tomorrow. I have changed my mind."

VINTON CERF: There are a number of reasons for regulating. The obvious one that comes up is abuse of monopoly power, but let me ask you, Nina, whether you sense other reasons for regulating, especially in this new information infrastructure environment, other than to avoid abuse of monopoly. Is there a public interest, a public good that needs to be protected, which is not necessarily subject to abuse by monopoly power?

CORNELL: That is a tough one. I sit wrestling in a very personal way with the whole issue of rural telecommunications. Those of you who know me know that I live 20 miles outside of a town that has 362 people and the nearest city of any size is at least a three-hour drive away. I happen to believe it would be a very bad idea to see rural America disconnected from the network, and I am concerned about that. At the same time I can tell you that having looked up close and personally at the "subsidy" mechanisms, and having discovered that their present value is about to be handed to US West as a gift as they walk away, I am not very convinced that we are doing a very good job of it. Maybe it would be better if entry were permitted.

I would love it if somebody wanted to do a voucher system so that it is not capitalized and given away, but I do not know the real answers to that. I am concerned about Eli Noam's third bubble, the universal service one [Figure 4A,B], because I am concerned about disconnecting—not me but everybody in the community around me—from the entire national network, which is, in fact, a real threat at the moment.

GEORGE TURIN: I need some help here. Let me start with a disclaimer. I am not now, nor have I ever been, an economist, so I really need some help. Everything I have heard from this panel, even from those who defend some aspect of regulation, seems to be in the direction that regulation will stop entities from doing things that they might do if the free market had total play, and those things that they might do might be bad, so regulation has some role. That seems a very reactive statement about regulation.

On the other hand, we heard in the first session this morning from several user communities, three out of four of which are going to need some real help. Now, here is where I need some help myself. Tomorrow's session concerns government investment. Is positive government investment the only way that those user communities are going to get help? Is there no way that regulation per se can act as a proactive force to help user communities that may be disenfranchised in some way—such as the library, education, and health care communities?

NOLL: Who would like to speak for using regulation to help health care?

CORNELL: Some of the kinds of concerns raised this morning were flat-out government: government pays for education, textbooks, and school buildings, and if pipes to the Internet are important, government should pay for that. But in the drive to price everything in telecommunications usage-sensitively versus flat rate is an issue that has some major implications, I think, for education, health care, and libraries. In that regard, there are some interests and some questions, and I do not think these are issues of subsidies. I was encouraged when Bridger Mitchell said he was not certain that flat-rate pricing was inefficient, because like one of the blind people trying to identify an elephant from a particular peculiar piece, I see measured usage often being used as a way to try to curb entry and to create barriers to use rather than being really efficient. I think there are some issues regarding flat-rate versus usage-sensitive pricing, particularly at the local level. It is a peculiar situation, particularly in the case of education and public libraries. We do view those as things we fund publicly, and so the issue is not really subsidy. The issue is funding, but I think, as I said, there is a different issue in regulation there.

CRANDALL: I do not know anything about libraries, but if you think about public education and the health care sector, you think of two institutions in our society that are starved for funds. I think what is wrong there is the incentive systems—there is no incentive to be efficient or to adapt to the right technology. I do not see any reason why government subsidies for infrastructure would improve their performance in any way.

NOLL: I agree, but in a certain sense the issue still is the reason that the educational system or the library system would choose to allocate as it has all of its previous budget increases. Recently there have been vast reductions in budgets for libraries, but the elementary and secondary education system budget has been growing in real terms for most of the post-war period but has by choice not been invested in telecommunications.

Now, the explanation for that could be, as Eli Noam put it, a critical mass problem. But it is hard to imagine a state the size of California or New York having a critical mass problem. In California, the state educational budget is determined by the state legislature. The funding is all centralized, and the state could in fact take 3 or 4 percent of the educational budget and invest it in telecommunications for educational systems, but it has chosen not to.

I think that before you get to George Turin's issue, you have to identify exactly why telecommunications hasn't been invested in for education and what the problem is for which we are trying to find a solution. In principle that is an enormous public-sector activity that does not choose to buy textbooks, let alone to invest in telecommunications. It chooses to do other things, and if the world is such that the politics of education selects against the delivery of information to students in favor of something else, it is exactly the same politics that will drive a regulatory institution. They are not separate. They are run by the same state legislature and governor. So what is it that regulation can solve that the state-centered budgeting of education cannot solve?

TURIN: So in short form, the answer is no.

NOLL: Yes.

LAURA BREEDEN: I am the executive director of FARNET, which is an association of Internet service providers. I am neither an economist nor a political scientist, but I did have to do this for a long time using some government money and some money that we got from clients, so I have a certain practical feel for these issues. I do not know what the right answer is in terms of regulation, but I can tell you that when you sit down to do your business plan or your budget for the year, what you are looking at is a huge expenditure on lines.

The cost of those lines for the long-haul sections has gone down and down. In fact, it does not much matter whether you are calling from Boston to Syracuse or Boston to San Francisco; it is still 22, 23, 24 cents a minute. The real cost is in the local loop facilities, which have not been deregulated. So when I hear someone ask if regulation is going to make a difference, I look at long distance and say, "Has regulation made a difference in the price of service?" I am one of the systems integrators. The world that I come from is the world of the systems integrator, where you buy the raw material, in this case the transport, from the phone company or from the bypass vendor and then you add value, the routers and Internet service, and so forth. So can regulation help? Deregulation can help. Deregulation might bring the cost of the local loop down so that you are not looking at $37,000 for an Internet connection, $20,000 of which is in the local loop, which is an artificially high cost. Maybe deregulation is a form of regulation. Maybe that is the answer.

NOLL: I am going to add one more item to the program and then we will go to discussion from the floor. Notice it was mentioned by both Tom Long and Eli Noam in their discussions that an important element of policy making is to figure out what the costs are, not only of the existing system but also of alternatives. To my knowledge by far the most comprehensive and best study of the costs of the telecommunications system under the current technology was performed by a group at the Rand Corporation led by Bridger Mitchell, so I have asked Bridger to tell us briefly the basic story of that study.

BRIDGER MITCHELL: I think Tom Long really raised one of the key questions by saying one of the fundamental issues is how to share a common cost, a cost that the telecommunications must incur in order to provide a whole range of services. Much of the debate in the regulatory commissions and the policy community has been confused; it has been about allocating that common cost among one or the other or several services. But the correct question in terms of a pure cross-subsidy is whether the prices are so low that the firm and the remainder of its customers would be better off by not supplying a particular group or a particular service. If the firm's actual net costs would go down or its net profit would go up by getting rid of the service, getting rid of the set of consumers, then that would be a cross-subsidy.

Several years ago, the California Public Utilities Commission joined forces with Pacific Bell and GTE in California and Rand to actually look at the technologies for delivering current-generation access and local exchange service. They put together a building block model, as it were, of the technology that is in use there and pretty much throughout the country and using data from California. Now, what came out of this? We were looking at the cost of adding subscribers to the existing network where the telephone system was already operating or adding additional usage by subscribers who were already on the network. And the basic findings that I would summarize here are that most of the costs of additional use or additional subscribers are fixed on a per-subscriber basis when one is looking at access to the network and use in the immediate local environment.

The variable costs that come out are primarily driven by the need to have additional capacity for peak or busy-hour utilization in order to provide and maintain quality of service. So one consequence of that is that flat-rate service for local calling, immediate local areas, does not look terribly inefficient when most of the costs of additional usage are primarily driven by peak-hour capacity. Another basic finding in that study is that the variation in costs across the state are driven primarily by the distance of the subscriber from the local switching center or the local wire center, and so rate averaging is discouraging entry in the high-cost, predominantly rural or suburban communities.

Finally, incremental cost, which is really the numbers we were looking at, is substantially below the average cost of the total plant if you were to build the local exchange from the ground up. That would suggest that on a statewide average, incremental costs are considerably below the $25-per-subscriber-per-month number that was being used in these filings.

NOLL: Just to put an important point on it, if you ask what is really subsidized, the bottom-line answer is, telephone services in small communities and rural areas, from the point of view of what the economist defines as a subsidy (which is actually paying less than the incremental cost of the service provided). So the really important point here regarding the debate about pricing and cost allocation is that there is a hunk of costs arising from the sort of economies of scale in the local telecommunications system and there is a fixed cost, so that the regulatory process becomes the mechanism whereby people fight out that battle in a legalistic and political sense about who is going to bear what cost. The outcome is that it is not the case that large numbers of people end up with subsidies.

It is the case that rural areas and small towns end up with subsidies, but everybody else is basically paying more than their incremental cost. There are economics we could then bring to bear concerning how you divide up that fixed cost most efficiently, but the problem is not really lots and lots of people getting subsidies. The problem is the fundamental conflict that was represented in the session between the technical and economic features of the industry to begin with and the political determination of how the cost of the system will be borne.

CHARLES JACKSON: Mr. Long, you stated, if I recall correctly, that the cost of the local loop should not be allocated just to basic service, and we have heard people talk about unbundling. Several questions here were about unbundling.

I live in Maryland. In Maryland a company called MFSI has applied to offer competitive local exchange service, at least in Maryland, to business and government. I will quote from one of their filings: "MFSI proposes that C&P [the local exchange carrier] be required to allocate the costs of providing dial-tone lines between two fundamental functions, links and ports, and to tariff separate unbundled rate elements for links and ports."

My first question, as a consumer advocate, is whether Maryland should approve or deny MFSI's request for such unbundling, and if they do unbundle, with or without your approval, whether all the cost of loops should be allocated to links and whether C&P should be allowed to geographically de-average link costs or should be required to maintain some sort of statewide average pricing of these things facing business or competitors?

THOMAS LONG: You will have to excuse me. I am not quite sure what MFSI meant.

NOLL: The idea being that somebody could get into the business of taking the C&P copper-wire loops but then connecting them to their own switch instead of the C&P switch. Or the alternative option.

LONG: I am afraid you have stumped me for now. I want to think about it.

HATFIELD: I would just add an anecdote. If you go back in the course of early history that we talked about, you find there were far more lines in the United States. One of the reasons we had rapid roll-out of telephone service is in part because farmers constructed their own lines and connected them to an existing switch in town. In this discussion, we have done a 360. We are back to the idea that maybe farmers should be able to construct their own lines and connect to an existing switch.

NOLL: I think the question about de-averaging is really crucial, because it not only is the loops versus ports story, it is that within the same town, some people are going to be 100 yards from their first switch and some are going to be 5 miles. Concerning Bob's point about the network architecture being not necessarily the most efficient, the only way to find out if the current arrangement is the most efficient is to have a relatively small number of very large switches in cities with lots of people connected a long way away from them, and then to charge the people who are 5 miles away ten times as much for telephone service as the people who are 100 yards away.

HARRIS: But by that, you do not necessarily mean they pay that out of their own pocket—have them face the price and decide?

NOLL: That is precisely the point. The reason it is a test is because the people who live 5 miles away and who face $100 a month for a telephone bill, right now in the current environment would have a very, very strong incentive to go for a cellular telephone instead because it would be cheaper.

WALTER BAER: Bob Harris spoke eloquently about the need for some federal preemption of state regulation. I would be interested in how the other panelists would comment on that, particularly regarding competitive entry. Also for Eli Noam's scheme: Eli, can that be implemented on a state-by-state basis, or does that require federal rules in implementation?

ELI NOAM: It would require federal principles and state implementation, which I think is basically the way federalism in the United States should function in the future or to some extent functions today. If you have redistributive systems or support systems on a state-by-state basis, presumably you would have some kind of a race-to-the-bottom competition among jurisdictions and all kinds of manipulations by carriers to shift costs and revenues. At the same time, if you had it totally centralized, you would have early implementation problems or you would not let New York or Wyoming go their different ways. So I think within the principles of federally set principles, you can permit local variations.

NOLL: Eli, address the following problem. I think this is the thrust of Walter Baer's question. Suppose the FCC takes the position that for state regulators that adopt the following state regulatory rules we will allow reintegration of the Bell operating companies. How would that work out exactly? Would you imagine that all 51 jurisdictions would adopt basically the same regulatory system for the local loop?

NOAM: I do not think they would. It is not clear to me whether they should. Take local competition. I think some of the states, New York and Illinois, for example, have been at the forefront of it and in a way have provided some of the models that have been implemented on the federal level either by the FCC or by a federal jurisdiction right now. That is, in a way, the way the system should work. You have a kind of basis for experimentation in the states, and if I were to fault the states it would actually be for not being innovative enough, for not using more of the competitive ability to be flexible and to have variations but rather congregating around generic kinds of resolutions to lowest-common-denominator policies. That is really the problem, not state experimentation.

Now, after a certain period for states to experiment, you can imagine a system in which the FCC or some other federal body will make this nationwide as long as you do not cut off the flexibility. I think it would be a real mistake.

CERF: It seems to me that one of the things that gets in the way of effective competition is lack of continuous competition. There are a lot of examples where you do not get this opportunity —in the cable companies, for example, since it is a franchise arrangement. If there is any competition at all, it is at the beginning of the franchise and then there is some period of time before which there might be any other competition. Some of the government telecommunication system services are like that. FTS-2000 is an example of an intermittent competition, it seems to me. (If I get this all wrong, the economists will fix me up, I am sure.)

But is there any way for us to assure that, as we enter into this new information infrastructure age, we are able to inject an opportunity for continuous competition so that consumers can change their minds and move from one supplier to another regularly? The first question is, Is that any good, is that useful, is that a helpful mechanism? and the second is, Is there anything getting in the way of doing that?

NOAM: It is a big question, but I would say the one way to deal with this is to make possible competitive entry that is less than end to end, for example, so that would mean a certain amount of unbundling of services for the telephone industry. In the context of the cable industry, it would mean the unbundling of the set-top converter-box-type technology from the network provision and a whole range of intermediate access issues so that entry can be partial rather than full. That is only a partial answer to your question.

CHARLES OLIVER: Eli Noam made the assertion that telecommunications is not price elastic and therefore it was okay to impose subsidy burdens on it. Does anyone else agree with that as a global assertion? Certainly we have seen the demand for access to interstate long-distance service more than double since the subscriber line charge was phased down. The implication is that basic local service is what should be bearing the subsidy burden and it should be cross-subsidizing long-distance service.

NOLL: Well, to defend Eli, he was not attempting to defend any particular cross-subsidization. What he was saying is that if a "philosophy king" who also happened to be an economist descended upon the earth and decided he wanted to raise X dollars in tax revenue, what would he tax? The answer is he would tax only the most inelastic demanded things, and basic access to the telephone network is about as inelastic a demand item as there is. That is all he meant.

OLIVER: So by that logic, if you could, you would tax oxygen and food?

NOLL: No. It was discovered at the time of the *Doomsday Book* that the right way to tax people is proportional to the heads they have.

Part 3

Public Investment in Telecomunications Infrastructure

Introduction to Part 3

David G. Messerschmitt

This session of the workshop discussed different mechanisms for increasing investment in infrastructure through direct or indirect public investment. Areas for investment or subsidy that were specifically considered were basic research, development, service and information providers, and end users.

A recent report from the Clinton administration, *National Information Infrastructure: Agenda for Action* (IITF, 1993), laid the groundwork by suggesting some areas for direct investment or subsidy. It proposed the basic thesis that carefully crafted government action can complement and enhance the efforts of the private sector, and assure the growth of an information infrastructure available to all Americans at reasonable cost. This thesis addresses two phases of evolution of a network infrastructure where free market mechanisms are inadequate, as identified by Eli Noam during Panel 2's discussion: the lack of critical mass early in its evolution (when the marginal utility to users does not justify the cost) and the desire for universal service in the mature phase (when the network is widely but not universally available, disadvantaging a subset of the population). Public investment or its close cousin user subsidy may be needed in the early phases to "jump start" the infrastructure, and cross-subsidy or user subsidy may be required later to extend the infrastructure to all users, particularly disadvantaged people or those in rural areas.

Four specific actions related to public investment or action to encourage the national information infrastructure (NII) are identified in the Clinton administration report:

1. Help the private sector develop and demonstrate technologies and applications. This most likely takes the form of government support of research or may take other forms such as tax credits or subsidy of private research.

2. Adjust federal procurement policies to provide incentives for the private sector to contribute to NII development. The question here is whether the government is in a position to choose the winning technologies.

3. Make the vast reservoir of government information available via the NII. Since the government at all levels has traditionally provided library service, to what extent should this extend to becoming an NII information provider?

4. Extend the universal service concept of the telephone network to the NII. Since the NII is likely still in the "critical mass" phase, it may be premature to consider this issue. Nevertheless, there are significant issues inherent in introducing the new NII into the

existing telecommunications environment, with its existing cross-subsidy structure and regulatory disincentives to the introduction of new technology.

Walter Baer gives an enlightening perspective on past government investments in telecommunications, beginning with the construction of the first telegraph system. Starting with the formation of Western Union and later the Bell system, most telecommunications infrastructure was constructed with private funds under the watchful eye of state and federal regulators. However, the government did support research and development of many telecommunications technologies for defense needs, and many of these later became important commercial successes, for example, microwave radio, computing, digital signal processing, satellites, and computer networking. Now that defense needs are giving way to concern for civilian economic development, one question is whether the government should continue to support technological development, and, if so, how it can best shift from a defense- to a commercial-based investment. Baer describes some ways that government is currently investing in telecommunications and other technologies:

- *Research and development tax credits.* A shortcoming of the current approach is that software tax credits are more difficult to obtain and yet are the lifeblood of the telecommunications industry;

- *Direct investment in research and development,* much of it subsumed under the High Performance Computing and Communications Initiative;

- *Networks and systems.* This includes a direct investment in networking infrastructure in the National Research and Education Network program. It also includes military systems such as the ARPANET and the Global Positioning System that are now making a transition to important commercial uses. It also includes investment in infrastructure for internal government use, such as FTS-2000, that has a substantial impact on commercial technologies because of the size of the procurements;

- *Subsidizing of user networking,* which is relatively infrequent but includes support for rural telephone networks through the Rural Electrification Administration telephone loan program;

- *Support of applications,* such as telemedicine, which is spread through a number of government agencies and is relatively small in the aggregate;

- *Subsidizing of users,* which is also rare but includes support for school and college connection to networks and support of public television stations;

- *Direct funding of agencies that regulate or set standards,* such as the Federal Communications Commission, the National Telecommunications and Information Administration, and the National Institute of Standards and Technology. These programs have tremendous leverage over private investments in telecommunications, especially in comparison to the size of government expenditures; and

- *Support for information and databases.* The government is the largest creator, collector, user, and disseminator of information.

Baer makes the important point that all federal investments in telecommunications infrastructure amount to between $0.5 billion and $2 billion per year (depending on how broadly you define infrastructure), which is tiny in comparison to the roughly $50 billion per year spent in the private sector. Since the government is not in a position to greatly increase these investments, direct government investment will remain tiny in comparison to private investment. Thus, the government will make the greatest impact if it focuses its investments in areas that will maximally leverage private investment. There are three important rationales for government investment where market mechanisms fail: redressing underinvestment in research and development, achieving critical mass in new technologies, and achieving equity in access. Given budgetary limitations, government can have a much greater impact through policies, standards, and regulations than it can through direct investments. The exception is research and experimentation, where government expenditures are a significant factor and are justified by shortcomings in market mechanisms.

Baer finishes by asserting that given these factors and the shift of government attention from military to economic and social purposes, the government should shift from the supply to the demand side, focusing more heavily on end-user applications and end-user support, rather than on networking infrastructure. It should also focus more on the diffusion, as opposed to just the development, of new applications. This may include some direct support of end-user groups, such as a voucher system. Baer also urges skepticism in accepting arguments that investment in telecommunications infrastructure will yield compensatory savings in administrative or other costs.

Bridger Mitchell discusses direct government construction of network facilities and subsidies to end users. He reiterates the issues relating to the critical mass problem in the early stages of a new network technology and the dissemination or universal service problem that arises as the network grows and matures. Most rationales for government subsidies of networks and users redress one or the other of these problems.

Charles Jackson considers government investment from the perspective of telecommunications carriers. He points out that the capital flows and usage charges are huge in relation to any credible government expenditures, and thus the government is unlikely to make any significant difference by direct investment in facilities or subsidies to users. Rather, the government should concentrate on the tremendous leverage it has in regulatory policy, for example, bringing the depreciation time of assets into line with the increasingly rapid obsolescence of the equipment.

William Gillis speaks from the perspective of an information supplier. He argues strongly for allowing marketplace mechanisms full rein, and against making prior assumptions about the information marketplace based on historical conceptions of the technology. Government should invest in the timely creation of standards for network interoperability and the establishment of testbed facilities but not the direct creation of services, where there is adequate private-sector activity. An exception is the creation of applications within the scope of government activities such as education. Gillis reiterates that applications never move from zero to a mass market quickly and that discussions of universal service in information services are therefore grossly premature. In the area of information services, he urges the government to make its vast databases available electronically but in the process to also reorganize the data and invest in the development of a uniform nomenclature to make it more accessible. The goal should not be simply to unleash runaway data, but rather to turn the data into easily accessible and useful information.

Robert Kahn addresses government investment in research and development. He begins by describing a conceptual framework of a generic network core, a demand-specific layer, and applications built on top of that (similar to the computer central processing unit, operating system, and applications). He emphasizes that applications are what matter to the end user but that the government has a critical role in coordinating and standardizing the core and demand-specific layers. It is extremely important that the core be designed in an "open system" fashion so that market mechanisms can freely define and deploy the applications without interfering with one another. The government should also ensure that there is a level playing field for all commercial participants. Kahn describes in some detail the conception and development of standards for the Internet as a premier example of how relatively small government investments in research and experimental testbeds can have a tremendously leveraged impact on the technology, economy, and society. He also gives a number of examples of companies (such as Sun Microsystems and Silicon Graphics) that have been the result of support by the Advanced Research Projects Agency to universities and another (Cray Computer) that was strongly influenced by government procurement and applications. His basic point is that government investment in basic research and precompetitive technology development can pay great dividends for the economy.

Laura Breeden, then of the Federation of American Research Networks (FARNET), speaks from the perspective of a systems integrator who repackages network services for nonprofit organizations. She emphasizes, as did earlier speakers, that the success or failure of the NII will be founded on the applications and the user friendliness and cost of the access devices, not the transport of the network. Also needed is a more active assessment program for determining the value of government investments and helping to guide future investments, rather than simply following the "religion" that the NII will be built and the benefits will be there. Finally, she appeals for greater scrutiny in the areas of massive federal investment, such as the defense and intelligence communities, as opposed to the relatively small scale of investments in research and educational networking. Can those investments be targeted for a greater commercial payoff?

The discussion following the panelists' presentation brings out a number of interesting issues:

- Questioners reiterate the importance of the applications and their value and friendliness to the end user, rather than the networking technology itself.

- The distinction between private and public investment in the telecommunications infrastructure is perhaps not as great as one might think, since virtually all taxpayers are also regulated ratepayers. Mistaken investments are subsidized by either the taxpayer or the ratepayer, which are one and the same.

- The NII will be used for accessing information services perhaps as much as for people-to-people communication. The question arises as to what universal service means in that context. What good does it do to get poor people or rural residents connected to the infrastructure if they cannot afford to access the information sources? In addition, an increasing portion of the cost is in access devices (e.g., computers) rather than in network access, and again this is an important obstacle to universal service.

- Government should consider and assess not only the economic aspects of the NII but also the impact on society. In the past, major new telecommunications infrastructures such as those for the telephone and television have had profound impacts, both positive and negative.

REFERENCE

Information Infrastructure Task Force (IITF). 1993. *National Information Infrastructure: An Agenda for Action.* Information Infrastructure Task Force, Washington, D.C., September 15.

Government Investment in Telecommunications Infrastructure

Walter S. Baer

INTRODUCTION

U.S. government investment in telecommunications infrastructure dates back at least to 1843, when Congress appropriated $30,000 to support construction of Samuel Morse's American Telegraph System between Baltimore and Washington, D.C. (Thompson, 1947, pp. 16-19). In the 150 years since then, the federal government has supported many other advances in telephony, terrestrial broadcasting, satellite communications, and data networking. But unlike investments in highways, mass transit, water supply and treatment, and related categories of physical infrastructure, telecommunications investments have been made largely by private-sector firms under regulations set by federal, state, and local authorities. Consequently, direct government investment in the U.S. telecommunications infrastructure has remained quite small compared to investments by telecommunications common carriers and other private firms.

In the past, federal investments in telecommunications have been driven principally by national security considerations and other direct government missions. Defense spending during World War II and throughout the Cold War period spurred the development of microwave and satellite transmission systems, computers for switching and network control, and new network concepts such as packet switching. Today, however, public concerns are shifting from military security toward U.S. economic competitiveness and societal needs in such areas as health care, education, and the delivery of public services.

The growing interest in telecommunications and information infrastructure reflects these economic and social policy concerns. Like other infrastructure elements, telecommunications and information networks directly and pervasively support both public- and private-sector activities. Their expanded use is generally believed to increase productivity and output, spur innovation, and lead to significant changes in organizational structure.[1] And they are changing rapidly.

The policy importance of technological advances in telecommunications and information is twofold. Not only do these advances promise even greater benefits to individuals and organizations who use them, they also break down the regulatory fences that governments have erected to keep industries separate. Technology is rapidly blurring the boundaries between the telephone, cable, broadcasting, and computer industries; between point-to-point communication and mass communication; and between communications and information services. These changes raise anew questions of how public- and private-sector responsibilities for infrastructure investment should be divided.

This paper discusses the issues surrounding public-sector investment in telecommunications infrastructure. It basically addresses the following questions: In what ways and for what reasons does the government invest directly in telecommunications infrastructure? Should these investments change in light of the rapidly advancing technology and increasing importance of telecom-

179

munications and information? The paper primarily focuses on federal investments in telecommunications infrastructure, recognizing that state and local governments play important roles as well (Mechling, 1993). Examples include state government investments in telephone and data networks, state university investments in local area networks, and local government investments in police, fire, and other communications networks and facilities.

This discussion of direct public investment complements the prior workshop session on how government influences private infrastructure investment through legislation, regulation, spectrum allocation and assignment, standards setting, and other "indirect" means. It should be emphasized that these indirect government activities have considerably more effect on total investment in telecommunications infrastructure than do direct government investments or subsidies.

A Note on Definitions

Defining the telecommunications infrastructure is tricky, particularly if the intent is to differentiate the "telecommunications infrastructure" from the "communications infrastructure" or the "information infrastructure." The 1991 infrastructure report from the National Telecommunications and Information Administration (NTIA) distinguishes between the "communications infrastructure which, broadly speaking, encompasses all of the facilities and instrumentalities engaged in delivering and disseminating information throughout the nation" and the more narrowly defined "telecommunications facilities, by which we mean facilities that permit point-to-point, two-way transmission of information of the user's choosing" (NTIA, 1991, pp. 13-14). The broader NTIA definition seems akin to the current "national information infrastructure" that includes point-to-point communications, mass media, the U.S. Postal Service and express delivery services, publishers and printers, and the motion picture and video industries.

This paper defines telecommunications infrastructure as encompassing all electronic communications, but not print. Unlike the NTIA definition, it includes both point-to-point and mass media such as cable television, since their technologies are swiftly converging and both can be used for the transmission of video, voice, and data services. The definition of infrastructure here also comprises four principal components: (1) the physical infrastructure, including both network and user facilities and their operating software; (2) applications software; (3) research and development on physical infrastructure and software; and (4) human capital investments relating to these other elements. This again is broader than the NTIA definition that includes only the physical infrastructure.

FEDERAL INVESTMENT IN
TELECOMMUNICATIONS INFRASTRUCTURE

The federal government subsidizes or directly invests in telecommunications infrastructure in a variety of ways, ranging from funding research and development (R&D), to building prototypes and operating systems, to supporting users who purchase telecommunications equipment and services in commercial markets. The principal categories of direct government support and subsidy are listed in Box 3 and described below.

**BOX 3 Categories of Federal Investment
in Telecommunications Infrastructure**

Tax Incentives

Direct Support of Research, Development, and Demonstrations
- Internet
- Other components of the High Performance Computing and Communications Initiative
- Defense research, development, testing, and evaluation
- Demonstration and prototype systems

Support of Telecommunications Networks and Systems
- Defense and space systems
- FTS-2000 and other government systems
- National Communications System
- Capital and operating subsidies

Development and Support of Applications

Support of Users
- Hardware and software purchases
- Operating subsidies
- Training

Other Types of Federal Support
- Support of regulation (e.g., Federal Communications Commission)
- Support of policymaking (e.g., National Telecommunications and Information Administration)
- Support of standards (e.g., National Institute of Standards and Technology)
- Databases and information

Tax Incentives

Since 1981, the federal government has subsidized technology investments in general through research and experimentation (R&E) tax credits, available to firms whose expenditures on laboratory or experimental R&D are above a base level.[2] Most communications equipment manufacturing firms and carriers benefit from R&E tax credits.

One controversial issue since these tax credits were first introduced is the extent to which software development qualifies as eligible R&E. Software is an increasingly important part of the telecommunications infrastructure, so that extending the range of software eligible for R&E tax credits would be a stimulus to infrastructure investments.

Investment tax credits were eliminated in the 1986 Tax Reform Act. Legislation to reenact them has [previously] been proposed to stimulate overall U.S. investment, but it seems unlikely to pass in the current Congress.

Direct Support of Research, Development, and Demonstrations

The administration budget for fiscal year 1994 (FY 94) included $72 billion for research and development, of which $30 billion (42 percent) was for nondefense purposes. Most of the nondefense R&D support for telecommunications, as well as $343 million from the Advanced Research Projects Agency (ARPA) of the U.S. Department of Defense, is now subsumed under the $1.1 billion multiagency High Performance Computing and Communications Initiative (HPCCI). The HPCCI includes five principal components:[3]

1. High-performance Computing Systems
   - Research for future generations of computing systems
   - System design tools
   - Advanced prototype systems
   - Evaluation of early systems

2. Advanced Software Technology and Algorithms
   - Software support for "Grand Challenges"
   - Software components and tools
   - Computational techniques
   - High-performance computing research centers

3. National Research and Education Network (NREN)
   - Internet
   - Gigabit research and development

4. Basic Research and Human Resources
   - Basic research
   - Research participation and training
   - Infrastructure
   - Education, training, and curriculum

5. Information Infrastructure Technology and Applications (IITA)
   - Information infrastructure services
   - Systems development and support environment
   - Intelligent interfaces
   - "National Challenges"

The NREN and the new IITA program are the two components most clearly directed toward advancing the nation's telecommunications infrastructure. However, the entire HPCCI can well be considered to be a federal investment in telecommunications and information infrastructure.

Beyond its contributions to the HPCCI, the Department of Defense invests heavily in advanced communications research, development, test, and evaluation (RDT&E). The FY 94 line-item budget for defense RDT&E for "intelligence and communications" totaled more than $5 billion. Although it is difficult to estimate how much of this directly related to the national telecommunications infrastructure, defense requirements have in the past often led to technology developments with commercial applications. Examples of such national security programs include much of the early work on digital signal processing, R&D on the ARPANET, and research by the National Security Agency on encryption in commercial telecommunications networks.

The federal government has occasionally supported prototypes and demonstrations of telecommunications systems intended for widespread commercial use, of which the Morse telegraph system is the earliest example. In the 1970s, the National Aeronautics and Space Administration (NASA) sponsored the ATS-6 satellite program that successfully demonstrated direct broadcasting of television and other communications services from space. The ATS-6 demonstration convinced a number of developing countries to invest in communications satellite systems, thus benefiting U.S. suppliers of satellites and ground stations (Cohen and Noll, 1991). It did not, however, directly influence the pace of investment in satellites for the U.S. telecommunications infrastructure.

In September 1993 NASA launched the Advanced Communications Technology Satellite (ACTS), which is a direct descendent of the earlier ATS program. As its name implies, ACTS is intended to demonstrate advanced technologies—principally high-data-rate switching and processing onboard the satellite. NASA project managers say that ACTS has directly influenced commercial infrastructure investments such as Motorola's Iridium system, which will provide satellite links to wireless communications users (NASA, 1993, pp. 1-2). Critics claim, however, that commercial technology had bypassed the $600 million ACTS program well before its launch (Broad, 1993).

Support of Telecommunications Networks and Systems

The Department of Defense and other national security agencies have traditionally built and operated their own telecommunications networks, ranging from secure military communications links to the ARPANET, which began as a low-cost computer-to-computer network, to the Global Positioning System (GPS), which sends satellite signals that give accurate location information worldwide. Some defense networks such as the ARPANET and the GPS have evolved to become part of the telecommunications infrastructure, but most remain separate systems dedicated solely to defense missions.

A few civilian agencies such as NASA build and operate extensive telecommunications networks of their own, but most other federal agency communications are coordinated by the General Services Administration (GSA). Since 1987, GSA has maintained an Information Technology Fund to procure telecommunications facilities and services for federal agencies. Its largest contract is known as FTS-2000, providing intercity voice, data, and video services. Managed by AT&T and Sprint, FTS-2000 is a software-controlled "virtual network" whose traffic is physically commingled with nonfederal communications. Whether or not the primarily leased services under FTS-2000 constitute federal "investment" in a technical sense, they certainly support investment in the telecommunications infrastructure by private carriers. Through the Information Technology Fund, GSA also procures local telecommunication services and supports information security and emergency management programs. In addition, the federal government maintains a National Communications System Office responsible for providing critical communications needs during emergencies (NCSO, 1993).

The history of the Internet illustrates how federal support of telecommunications networks can change as the systems evolve (Hart et al., 1992). The original concept of distributed computer/communication links was based on research in the early 1960s supported by the U.S. Air Force (Baran, 1964).[4] In 1969 ARPA built the ARPANET, the first computer-to-computer network based on this "packet-switching" concept. The ARPANET was a fully government-funded network linking the Department of Defense and its contractors. By the early 1980s the ARPANET's success led to proposals to develop similar networks for nondefense uses. The National Science Foundation thus funded its own NSFNET to link its supercomputer sites and other major U.S. centers of computing research. Other regional and local networks quickly developed, linked by the NSFNET and other "backbone" networks and connected to many sites and networks around the world. Today, the Internet comprises tens of thousands of interconnected, interoperable computer networks. Federal direct support represents only about 10 percent of total Internet costs in the United States (Hart et al., 1992, p. 686).[5] As of July 1993, more than half of the 46,782 registered networks connected to the Internet were commercial.[6]

Whether federal agencies should continue to provide network subsidies as the Internet becomes part of the national telecommunications infrastructure has emerged as a key policy issue.

The federal government generally does not provide capital or operating subsidies for telecommunications, except for special congressionally mandated purposes. Since 1949, the Rural Electrification Administration (REA) has offered low-interest loans and other subsidies to support the expansion of telecommunications to rural areas. And the National Telecommunications and Information Administration has awarded planning and construction grants, primarily for public broadcasting and educational television, through its Public Telecommunications Facilities Program.

NTIA is now expanding its grants program to include networking pilot projects, sometimes described as network "on ramps." These grants provide matching funds to schools, libraries, state and local governments, and other nonprofit organizations to purchase the equipment needed to connect to computer networks such as the Internet.

Development and Support of Applications

The federal government has often funded innovative telecommunications applications for defense, education, health care, criminal justice, agricultural extension, and other areas of federal agency concern.[7] While past applications have successfully used existing voice, video, and low-speed data facilities, the emerging high-speed data networks may make many new applications possible.[8]

To promote such applications, the administration requested $96 million in FY 1994 for the new Information Infrastructure Technology and Applications component of the HPCCI; $156 million was eventually allocated, covering a larger number of agencies than the request.[9] This effort "will develop and apply high-performance computing and high-speed networking technologies for use in the fields of health care, education, libraries, manufacturing, and provision of government services" (IITF, 1993, p. 9).

Even more difficult than developing new applications, however, is helping users adopt them and adapt them to their own circumstances. The innovation literature is replete with examples of how private firms and public-sector agencies often lag in utilizing new technologies or applications that have been demonstrated elsewhere. As a consequence, some government programs aim to speed adoption and diffusion through technology transfer or extension services. Intensive use of telecommunications and information networks is of growing importance for such diffusion-oriented programs.

Support of Users

Proponents of an expanded NREN propose increased government support to help bring schools, libraries, and other new users onto the network. The administration's program to expand the NTIA networking grants is a clear step in this direction. Legislation proposed in 1993 would have established within the National Science Foundation (NSF) a "Connections Program" to "(1) foster the creation of local networks in communities which will connect institutions of higher education, elementary and secondary schools, libraries, and State and local governments to each other; and (2) provide for connection of such local networks to the Internet."[10] The bill also would have authorized federal funds for training new users, which is no less important than purchasing the hardware and software necessary to connect them.

Up to now there have been no categorical federal programs to support telecommunications users. Recipients of federal research grants and contracts may cover some equipment, service, and training costs under their awards; and their institutions receive indirect costs that they can use to pay for telecommunications equipment and services. Federal laws may require telecommunications equipment suppliers and carriers to provide specific features or services, such as closed-caption decoders in television receivers[11] or "telecommunications relay services" that let hearing-impaired individuals communicate over the telephone network.[12] The FCC also offers matching funds to states that adopt "Lifeline" or other assistance programs to low-income telephone subscribers. But the specifics of user subsidies are generally left to state public utility commissions and state and local government welfare programs.[13]

Other Types of Federal Support

The federal government supports the telecommunications infrastructure in a variety of other ways as well. Most directly, federal support of policymaking, regulation, spectrum allocation, and assignment by the FCC and NTIA sets the ground rules that govern private-sector investments in the telecommunications infrastructure. These agencies as well as NIST support standards-setting activities in telecommunications. And government agencies through their own R&D and procurement sometimes set de facto standards, such as the TCP/IP protocols developed by ARPA and now used by the Internet.

As "the world's largest creator, collector, user and disseminator of information" (OMB, 1992, p. III-5) the federal government provides information and databases that flow over the telecommunications infrastructure. Recent revisions of the government's information management policies encourage electronic dissemination (OMB, 1993), and the National Technical Information Service, the U.S. Government Printing Office, and the Library of Congress, among other agencies, are developing ways to make more of their information accessible in electronic form. Although not direct investments in the telecommunications infrastructure itself, these efforts support the infrastructure and make it more valuable to other users.

SCALE OF FEDERAL INVESTMENT IN INFRASTRUCTURE

Federal investments in telecommunications infrastructure are not as well documented as are investments in other physical infrastructure categories. Table 1 presents 1994 federal budget requests totaling $49 billion for infrastructure investments in highways, rail and mass transit, aviation, water transportation, and water supply and treatment. Data published by the Congressional

TABLE 1 Federal Investment in Nontelecommunications Infrastructure

Category	Budget Authority ($M)	
	1993 Estimate	1994 Budget
Highways	21,439	21,321
Rail and mass transit	5,802	5,738
Aviation	10,466	10,863
Water transportation	8,639	8,724
Water supply and wastewater treatment	4,408	2,338
TOTAL	50,754	48,984

SOURCE: Congressional Budget Office (1993), Table 4, pp. 16-17.

Budget Office show that state and local governments invested more than twice as much in infrastructure as did the federal government during the 1980s.[14]

Data available in 1993 on federal investments in telecommunications infrastructure are shown in Table 2. As noted above, the largest direct investments have been for R&D through the HPCCI. The budget for defense RDT&E for intelligence and communications is five times greater than the HPCCI budget, but most of this cannot be linked directly to the national telecommunications infrastructure. However, the new Technology Reinvestment Program (TRP) of the Department of Defense to develop dual-use technologies [could have made] some contributions to infrastructure. The TRP identified information infrastructure, including network architecture, wireless communications, software development, and heterogeneous databases, as the first of eleven key dual-use technologies for funding (ARPA, 1993, p. A-1). TRP received $472 million in fiscal year 1993, with $600 million requested for 1994. [FY 1995 appropriations were recised.]

Under networks and facilities, GSA's Information Technology Fund had a proposed FY 94 budget of $1.3 billion, of which $834 million was for lease or purchase of interexchange and local telecommunications services and equipment. The direct subsidy component of the REA telephone loan program varies from year to year but is less than 15 percent of the total loan portfolio.

Federal budget requests for telecommunications applications and user support totaled about $230 million. These include programs in NTIA, REA, the Department of Education, and the Department of Health and Human Services, as well as the two initiatives for information infrastructure applications and networking pilot projects. Direct federal spending on telecommunications regulation, policymaking, and standards through the FCC, NTIA, and NIST amounted to about $160 million.

A narrow definition of direct federal investment in telecommunications infrastructure—including only the NREN, ACTS, the National Communications System, the REA telephone loan subsidies, the NTIA Public Telecommunication Facilities Program and networking grants, the IITA program, and other federal agency telecommunications applications programs—gives a 1994 total of about $530 million in proposed budget authority. Broadening the definition to include the entire HPCCI, the investment portion of the GSA Information Technology Fund, the small fraction of the TRP and Defense RDT&E (outside the HPCCI) that contributes to the civilian telecommunications infrastructure, and federal funding for telecommunications regulation, policymaking, and standards setting increases the total to perhaps $1.8 billion to $2 billion. This figure is still small compared to the $49 billion the federal government budgeted in 1994 for other physical infrastructure and to the $50 billion estimate for annual private investment in U.S. telecommunications infrastructure,[15] but it represented a significant increase over federal investment in prior years and signified the importance the Clinton administration has placed on advancing the telecommunications infrastructure.

RATIONALES FOR DIRECT FEDERAL SUPPORT

Given the large and growing private investments in U.S. telecommunications infrastructure, why should the federal government directly fund or subsidize additional investment? Can we not rely on the private sector to make better investment decisions than the government? The basic response is that in some cases the net benefits to society from infrastructure investment will exceed the benefits that individuals and firms can appropriate to themselves. This is the classic public goods argument for government support of highways, airports, and water and sewer systems. Some telecommunications investments may share these public goods characteristics.

TABLE 2 Federal Investment in Telecommunications Infrastructure

Category	Budget Authority ($M)		
	1992 Actual	1993 Estimate	1994 Budget
Research and Development			
National Research and Education Network	92	114	171
Other High Performance Computing and Communications Initiative components (except Applications)	563	681	829
National Aeronautics and Space Administration ACTS program	11	12	9
Defense research, development, training and education— intelligence and communications	4,584	4,910	5,168
Defense Technology Reinvestment Program		472	600
Networks and Facilities			
General Services Administration Information Technology Fund			
• Intercity services	554	589	508
• Local telecommunications	261	324	326
• Information security and emergency management	34	35	36
National Communication System	52	67	73
Rural Electrification Administration (REA) telephone loan subsidy (total direct telephone loans)	34 (240)	44 (344)	45 (359)
Applications			
Information Infrastructure Technology and Applications		47	96
Dept. of Education Star Schools Program	18	23	27
Dept. of Health and Human Services Medical Library Assistance	21	21	23
REA educational and medical links	5	5	10
Support of Users			
National Telecommunications and Information Administration (NTIA) public telecommunications facilities	20	20	21
• Information infrastructure networking pilot projects			54
Other Support			
Federal Communications Commission operations	126	141	130
NTIA operations	18	20	22
National Institute of Standards and Technology Computer Systems Laboratory	12	12	13

SOURCE: *Budget of the United States Government, Fiscal Year 1994;* other U.S. government documents.

The general public goods rationale can further be divided into four principal arguments for federal investments in telecommunications:

1. *Serving government missions.* This rationale is used especially to justify federal investment in R&D and advanced communications systems in defense, space, and intelligence operations. A related argument states that the federal government should directly support telecommunications infrastructure that provides net benefits to education, improved state and local government operations, and other activities that have public goods aspects, even when the federal government is not the direct provider of these goods or services.

2. *Balancing private underinvestment in R&D.* New knowledge is a true public good, expensive to generate but generally cheap to replicate, and consequently yields social returns much higher than private returns (CBO, 1991, p. 75; 1993a). Private firms and individuals will thus consistently invest less in R&D than is socially optimal. This argument lends particularly strong support for government funding of basic research but also is used to justify government support of "precompetitive" technologies and prototypes such as the gigabit network testbeds under the NREN program. It is also the underlying rationale behind the R&E tax credit.

3. *Redressing other private market failures.* Beyond underinvesting in R&D, private markets may fail to give the right signals for optimal investments in telecommunications for such reasons as:

• Economies of scale or scope, which characterize most infrastructure investments whether they be highways, electricity distribution systems, or telecommunications distribution networks.[16] The trend toward digital communications reinforces economies of scope, since a digital network can efficiently intermix voice, video, and data services and provide customized "virtual networks" to users. Economies of scale and scope constitute a principal rationale for government regulation of telecommunications services;

• Network externalities, which are more specific to telecommunications, since the value of the network to each subscriber increases as others connect to it;

• Uncertain demand for new services, which may lead telecommunications carriers and other suppliers to underinvest and thus support government efforts to aggregate demand or subsidize costs in order to seed the market;

• Lack of information among users that hinders them from signing up for or using telecommunications services. This argument is often made to justify government subsidies to public-sector organizations, small businesses, and individuals; and

• High transaction costs for users, often related to lack of information.

4. *Assuring equitable prices and access.* A clear purpose of government is to make sure that all citizens have fair access to and use of essential infrastructure. In the United States, issues of fairness in access to and cost of telecommunications services are generally handled within the regulatory process. Under the Communications Act of 1934 and subsequent legislation, the FCC sets the basic rules for common carrier, broadcast, and cable services. State and local government agencies then regulate prices of some services, often providing lower rates to certain groups of subscribers through cross-subsidies from other subscribers.

Federal direct investment has played a relatively minor role in fostering equity—one exception being the REA-subsidized loans for rural telecommunications systems. But concern about the new technologies exacerbating the gap between the "information rich" and "information poor" leads some to propose direct federal programs to provide Internet connections and other advanced services to disadvantaged groups.

Table 3 shows how these rationales relate to the categories of direct federal support discussed above. Of course, the conceptual basis of an argument for federal support does not mean that such support is appropriate or sensible. Many past federal investments in telecommunications to serve agency missions, for example, have proved short sighted.[17] Where infrastructure investment comes predominantly from the private sector, the burden falls on the government to justify its direct involvement. A recent forum by the Organization for Economic Cooperation and Development (OECD) on "Infrastructure Policies for the 1990s" concluded:

> In order to exploit more fully the scope for making greater use of private capital and building infrastructure, government and the private sector need to explore new and flexible ways of cooperating with each other—including the possibility of mixed financing—and establish criteria for the allocation of roles between public and private resources . . . . If a project's rate of return is perceived to be sufficient to attract private sector capital, then the project should be left to the private sector to construct. . . . Where a project's profitability is a borderline case and/or social equity considerations are involved, then the government should offer incentives to attract private capital . . . . Only where profitability is clearly negative but the social rate of return high should the project remain in the public domain (OECD, 1993, p. 6).

TABLE 3 Rationales for Direct Federal Investment in Telecommunications Infrastructure

Category of Federal Investment	Rationale for Investment				
	Federal Government Missions	Non-federal missions	Under-investment in R&D	Other Market Failures	Equity
Tax incentives			x		
Research and development	x		x		
Networks and systems	x			x	x
Applications	x	x		x	
Support of users		x		x	x
Other federal support Regulation Policymaking Standards Databases and information	x			x	x

DISCUSSION

Direct federal support of telecommunications infrastructure has primarily been on the supply side: support of R&D, testbeds and prototypes, and operating networks and systems. These supply-side initiatives work best when the objective is to serve government missions or bolster private-sector underinvestment in R&D. Underinvestment in research may be of particular concern in a time of industry restructuring and uncertainty such as that facing the telecommunications industry today.[18] When the objective is to improve economic performance by reducing market failures, demand-side initiatives seem more suitable. These can include stimulating adoption and diffusion of applications and innovative services or supporting users. When equity considerations dominate, legislative mandates (e.g., requiring carriers to provide services for the hearing impaired) or regulatory approaches may be preferred.

Past studies have concluded that the U.S. government is not very good at picking commercial winners in telecommunications or other advanced technologies (Cohen and Noll, 1991; Baer et al., 1977). Some allege that foreign governments have been more successful in building advanced telecommunications infrastructures, citing Japanese plans to deploy fiber optics to the home, and both European and Japanese investments in integrated services digital networks (ISDNs). Under close scrutiny, however, these examples are not very persuasive. Japanese plans for installing fiber to the home have been regularly deferred, so that the United States and Japan now seem on roughly comparable schedules. And the "ISDN gap" between the United States and Europe and Japan results largely from the ready availability of less expensive digital lines in the United States (OTA, 1993a, p. 183; Flamm and Weingarten, 1993).

Market feedback becomes essential when direct government support extends beyond the R&D stage. The lessons from past government efforts to demonstrate new technologies are clear (Baer et al., 1977):

- Involve industry and users in the project design and at all subsequent stages.
- Require significant cost sharing from participants and others who stand to gain if the demonstration succeeds.
- Have an explicit plan to hand off the technology to the private sector.
- Keep the demonstration focused on meeting its technical and economic objectives; insofar as possible, shelter it from capture by political constituencies.
- Avoid rigid schedules or tight time constraints.

These guidelines pertain to gigabit network demonstrations and testbeds under the NREN program as much as to past government-funded demonstrations in transportation or energy.

The federal government has opportunities to restructure its own activities in support of advancing the telecommunications infrastructure. The decision under FTS-2000 to lease virtual networks from private carriers rather than build separate government facilities is one excellent recent example. Others include:

- Making Internet services available through the FTS-2000 system. Assuring interoperability between FTS-2000 and the Internet could improve federal agency operations and enhance public access to government information and services (OTA, 1993b, p. 70).

- Accelerating efforts to make federal information and databases available on the Internet and other electronic networks. This is itself an important topic of current policy discussion, with controversies over developing directories and other tools to make access easier, appropriate pricing, potential competition with private-sector information providers, appropriate use of the Freedom of Information Act, privacy considerations, and other issues that are beyond the scope of this paper. Nevertheless, federal agencies should be encouraged to provide the public with greater electronic access to government-produced information.

• Extending federal technology transfer and technology extension services to make full use of telecommunications networks. In some cases this may mean the redesign of technology programs to emphasize demand-pull from users rather than technology-push from federal agencies.

• Expanding federal agency use of electronic data interchange (EDI) for forms and standard documents. The Commerce at Light Speed system originated by the Department of Defense is a good example of how EDI systems can lower costs and increase efficiency.

These and other efforts to improve government services and operations by using the telecommunications infrastructure more effectively have been a major theme of the administration's new program to "reinvent government" (Gore, 1993).

Federal support can also help develop telecommunications applications in education, health care, library access to information, and other ways to improve state and local government services; and it can help new users adopt these applications. The new programs for information infrastructure applications and networking pilot projects are cases in point. But one should recognize that the federal government is a relatively small player in these arenas. The federal role should be to support applications development and diffusion, not to provide permanent operating subsidies that could become expensive new entitlements. Steady-state funding of telecommunications applications should ultimately rest with the public- and private-sector users who find value in them.

Another way the federal government can directly advance the domestic telecommunications infrastructure is to reallocate spectrum from government to nongovernment use. According to legislation introduced in 1990, "the [federal] government currently reserves for its own use, or has priority of access to, approximately 40 percent of the spectrum that is assigned pursuant to the Communications Act of 1934."[19] Much of this spectrum was initially allocated to the federal government for national security purposes, but advances in technology and burgeoning commercial demand for wireless services suggest revisiting that decision. Reallocating underutilized government spectrum can greatly expand the infrastructure's capacity to provide cellular telephone, paging, and the next generation of wireless personal communications services. By auctioning spectrum licenses, it can also bring significant new revenues to the federal government. Congress began this process by authorizing the reallocation of not less than 200 megahertz as part of the Omnibus Budget Reconciliation Act of 1993.[20]

Beyond direct investment, the federal government's most important role remains that of sustaining an innovative, competitive telecommunications infrastructure provided by private-sector firms. Government's responsibilities are to maintain a sound economic climate for private investment and a regulatory framework that encourages fair and open competition among equipment and service providers. We no longer think of the telecommunications infrastructure as dominated by a single network that provides all services to all users. Rather, the infrastructure includes multiple networks with different functions, capabilities, and patterns of ownership and use. Assuring interconnection and interoperability of these networks is an important role of government at all levels, so that society can gain the maximum value from both public- and private-sector investments in telecommunications infrastructure.

REFERENCES

Advanced Research Projects Agency (ARPA). 1993. "Technology Reinvestment Project: Program Information Package for Defense Technology Conversion, Reinvestment, and Transition Assistance," Arlington, Va., March 10.

Akselson, Sigmund, Arne Ketil Eldsvik, and Trine Folkow. 1993. "Telemedicine and ISDN," *IEEE Communications Magazine* (January):46-51.

Baer, Walter S., Leland L. Johnson, and Edward W. Merrow. 1977. "Government-sponsored Demonstrations of New Technologies," *Science* 196(May 27):950-957.

Baran, Paul. 1964. *On Distributed Communications: Summary Overview*, RM-3767-PR. RAND Corporation, Santa Monica, Calif., August.

Broad, William J. 1993. "Satellite a White Elephant, Some Say," *New York Times*, July 20, p. C1.

Cohen, Linda R., and Roger G. Noll. 1991. "The Applications Technology Satellite Program," pp. 165-166 in *The Technology Pork Barrel*. Brookings Institution, Washington, D.C.

Committee on Information and Communication (CIC), National Science and Technology Council. 1994. *High Performance Computing and Communications: Technology for the National Information Infrastructure*. Office of Science and Technology Policy, Washington, D.C.

Congressional Budget Office (CBO), U.S. Congress. 1991. *How Federal Spending for Infrastructure and Other Public Investments Affects the Economy*. Congressional Budget Office, Washington, D.C., July.

Congressional Budget Office (CBO). 1993a. *A Review of Edwin Mansfield's Estimate of the Rate of Return from Academic Research and Its Relevance to the Federal Budget Process*. Congressional Budget Office, Washington, D.C., April.

Congressional Budget Office (CBO). 1993b. *Updating Trends in Public Infrastructure Spending and Analyzing the President's Proposals for Infrastructure Spending from 1994 to 1998*. Congressional Budget Office, Washington, D.C., August.

Egan, Bruce L., and Steven S. Wildman. 1992. "Investing in the Telecommunications Infrastructure: Economics and Policy Considerations," pp. 19-54 in *A National Information Network: Annual Review of the Institute for Information Studies*. Institute for Information Studies, Queenstown, Md.

Federal Coordinating Council for Science, Engineering, and Technology (FCCSET), Office of Science and Technology Policy. 1992. *Grand Challenges 1993: High Performance Computing and Communications, The FY 1993 U.S. Research and Development Program*. Committee on Physical, Mathematical, and Engineering Sciences, Office of Science and Technology Policy, Washington, D.C.

Federal Coordinating Council for Science, Engineering, and Technology (FCCSET), Office of Science and Technology Policy. 1994. *High Performance Computing and Communications: Toward a National Information Infrastructure*. Committee on Physical, Mathematical, and Engineering Sciences, Office of Science and Technology Policy, Washington, D.C.

Flamm, Kenneth, and Fred W. Weingarten. 1993. "Final Report to the National Science Foundation: Participation in International Study of High Performance Computing." National Science Foundation, Washington, D.C.

Gore, Jr., Albert. 1993. *From Red Tape to Results: Creating a Government That Works Better & Costs Less: Reengineering Through Information Technology, Accompanying Report of the National Performance Review*. U.S. Government Printing Office, Washington, D.C., September.

Hart, Jeffrey A., Robert R. Reed, and Francois Bar. 1992. "The Building of the Internet," *Telecommunications Policy* (November):666-689.

Information Infrastructure Task Force (IITF). 1993. *The National Information Infrastructure: Agenda for Action.* Information Infrastructure Task Force, Washington, D.C., September 15.

Johnson, Leland L. 1988. *Telephone Assistance Programs for Low-Income Households.* RAND Corporation, Santa Monica, Calif., February.

Mechling, Jerry. 1993. "A State-level View of Information Infrastructure: Aligning Process and Substance," pp. 31-45 in *Building Information Infrastructure,* Brian Kahin, ed. McGraw-Hill, New York.

National Aeronautics and Space Administration (NASA). 1993. "ACTS Technology Being Used by Industry," *ACTS Quarterly.* NASA Lewis Research Center, Cleveland, Ohio, August.

National Communications System Office (NCSO). 1993. *FY92 National Communications System's Annual Report,* NCS 2653/6. National Communications System Office, Washington, D.C.

National Science Board (NSB). 1993. *Science and Engineering Indicators,* NSB 93-1. U.S. Government Printing Office, Washington, D.C.

National Telecommunications and Information Administration (NTIA). 1991. *The NTIA Infrastructure Report: Telecommunications in the Age of Information.* U.S. Department of Commerce, Washington, D.C., October.

Office of Management and Budget (OMB). 1992. *Information Resources Plan of the Federal Government.* Office of Management and Budget, Washington, D.C., November.

Office of Management and Budget (OMB). 1993. *Management of Federal Information Resources,* Circular No. A-130. Office of Management and Budget, Washington, D.C., revised June 25.

Office of Technology Assessment (OTA), U.S. Congress. 1989. *Linking for Learning: A New Course for Education.* Office of Technology Assessment, Washington D.C., November.

Office of Technology Assessment (OTA), U.S. Congress. 1990. "Communication and Comparative Advantage in the Business Sector," pp. 107-142 in *Critical Connections: Communication for the Future.* U.S. Government Printing Office, Washington, D.C., January.

Office of Technology Assessment (OTA), U.S. Congress. 1991. *Rural America at the Crossroads: Networking for the Future.* Office of Technology Assessment, Washington D.C., April.

Office of Technology Assessment (OTA), U.S. Congress. 1993a. *U.S. Telecommunications Services in European Markets.* Office of Technology Assessment, Washington, D.C., August.

Office of Technology Assessment (OTA), U.S. Congress. 1993b. *Making Government Work: Electronic Delivery of Federal Services.* Office of Technology Assessment, Washington, D.C., September.

Organization for Economic Cooperation and Development (OECD). 1993. *Infrastructure Policies for the 1990s.* Organization for Economic Cooperation and Development, Paris.

Thompson, Robert L. 1947. *Wiring a Continent: The History of the Telegraph Industry in the United States, 1832-1866.* Princeton University Press, Princeton, N.J.

NOTES

1. Not everyone agrees with this assessment. For a discussion of relevant studies, see Egan and Wildman (1992) and OTA (1989).

2. Congress has extended the R&E tax credits for three years, and the administration has sought to make them permanent. See *Technology for America's Economic Growth, A New Direction to Build Economic Strength*, Washington, D.C., The White House, February 22, 1993, p. 27; *Technology for Economic Growth: President's Progress Report*, Washington, D.C., The White House, November 1993, pp. 23-24.

3. Gibbons, John H., testimony before the Committee on Science, Space, and Technology, U.S. House of Representatives, April 27, 1993. The HPCCI is described in FCCSET (1992).

4. Ten additional volumes detailing Baran's concept of distributed communications were also published by RAND in August 1964.

5. A good portion of the remaining 90 percent of costs is paid for by universities that receive some payments for indirect costs from the federal government, and by state and local governments.

6. Presentation by Tony Rutkowski at INET '93, International Networking Conference, San Francisco, California, August 1993.

7. See, for example, OTA (1989, 1991) and Akselson et al. (1993).

8. See FCCSET (1993), pp. 41-65.

9. See FCCSET (1994), p. 25, and CIC (1994), p. 50.

10. U.S. Congress, *High Performance Computing and High Speed Networking Applications Act of 1993*, H.R. 1757, April 21, 1993.

11. U.S. Congress, *Television Decoder Circuitry Act of 1990*, PL 101-431.

12. U.S. Congress, *Americans with Disabilities Act of 1990*, PL 101-336, July 26, 1990.

13. For example, see Johnson (1988).

14. See CBO (1993b), Table 1, p. 3.

15. Pepper, Robert, testimony before the Subcommittee on Technology, Environment, and Aviation, Committee on Science, Space, and Technology, U.S. House of Representatives, Washington, D.C., March 23, 1993. Pepper estimates that the $50 billion is "split almost evenly between network equipment and customer premises equipment." Some would argue that investment by regulated common carriers constitutes "public investment" in telecommunications infrastructure, noting that *The NTIA Infrastructure Report* (NTIA, 1991) compares U.S. common carrier investments with investments by government-owned carriers in other countries (see Table 5.2). However, this paper contends that investments by regulated private carriers in the United States (and increasingly in other countries as a result of privatization of government-owned telecommunications entities) are qualitatively different from direct government investments and should be clearly separated.

16. Wired distribution networks, including coaxial cable and fiber optic systems, show economies of scale; but wireless networks typically do not. Whether scale economies apply at the level of telecommunications services, when labor and other costs are factored in, remains an open question. P. Srinagesh, Bell Communications Research, private communication.

17. The rapid obsolescence of telecommunications facilities and equipment was a principal reason that the federal government turned to leasing and purchasing services from private firms under FTS-2000.

18. The most recent data collected by the National Science Foundation on industry-supported R&D in the communication equipment sector (SIC 366) show a small decline in constant dollars—about 1 percent per year—over the seven years following the AT&T divestiture in 1984 (NSB, 1993, Appendix Tables 4-1, 4-32, and 4-33).

19. U.S. Congress, *Emerging Telecommunications Technologies Act of 1990*, H.R. 2965.

20. Title VI, *Omnibus Budget Reconciliation Act of 1993*, August 10, 1993.

Federal Investment Through Subsidies:
Pros and Cons

Bridger M. Mitchell

Subsidies for public programs have an appealing ring to many enthusiasts of the information infrastructure—until one confronts the need to finance them! The "shadow price" of a subsidy dollar today is high. Additional federal spending for telecommunications investment requires increased taxes, a larger deficit, or cuts in other government programs—each is a measure of the opportunity cost of a greater subsidy. Efficiency in public budgeting would adjust government spending across diverse programs to equate the marginal benefit per dollar spent.

In this paper I consider the benefits that federal subsidies could yield by examining the types of market failures that may cause private investment to fall short of socially desirable levels. I then discuss policy instruments for directing public funding to the industry, consider the characteristics of supply-side and demand-side subsidies, and conclude with a highly stylized view of how subsidies may promote network growth.

MARKET FAILURES

Market failures can cause investment and consumption decisions reached in private markets to be economically inefficient. In such markets public tax and subsidy policy can potentially improve the performance of the economy. In the telecommunications sector three types of market failures can arise.

First, in a communications network existing subscribers place greater value on their network connection as more users join the network. The consequence of this "network externality" is that the social value of enlarging the telecommunications network by one user exceeds the private value expressed in a potential subscriber's willingness to pay (his/her maximum demand price). To a large extent, a single network supplier is able to internalize this effect by offering promotional prices and volume discounts that are recovered by increased revenues from a larger future network. However, if there are several competing suppliers, they may be unable to appropriate the full gains from eventual network expansion. The result is a failure of the market to reach its optimal size.

Second, the spread of telecommunications services is restricted by the limited income of some consumers. Social goals and principles of equity may then establish "universal service" as an objective, one that calls for expanding service beyond the point implied by the network externality.

A third failing of markets occurs when the commodity is a public good that, once produced, can be consumed by any number of users without additional cost. Information per se is a good that has zero marginal cost when an additional user consumes it. Many sources of information are consumed by using telecommunications services to perform search, access, and retrieval services at a distance. As a result of this complementarity of information and telecommunications, the derived (net) demand for information resources is increased as the costs of effective access (telecommunications and information processing) fall.

INSTRUMENTS FOR GOVERNMENTAL FUNDING

Governments have several classes of policy instruments with which to transfer resources into the telecommunications sector and to specific services.

Price discrimination and internal (cross-) subsidies permit revenues to be transferred from profitable, high-value services to other services. Public regulation has traditionally encouraged monopoly networks to set selected prices well above incremental costs and maintain low prices for favored services. Such rates will be unsustainable in competitive markets. In the telephone network, business subscribers and long-distance callers have been charged high rates in order to benefit residential users and local calling. However, major price discrimination requires that the supplier have market power. As more markets are contested by new entrants, frequently with new technology, the scope for "cross-subsidization" is diminishing rapidly.

Direct governmental subsidies to the telecommunications sector include research grants, demonstration projects, matching funds, state assistance programs, and low-interest loans.

Government provides in-kind assistance in the form of discounts or preferential access to public resources. Thus, the radio frequency spectrum has traditionally been transferred to private suppliers for only nominal licensing fees. The many inefficiencies that this allocative mechanism has created have long been understood, including the wasted resources expended by contenders for licenses that may largely dissipate any intended assistance.

Finally, through broad-based tax preferences for investment and research and development expenditures, government encourages higher levels of activities in sectors, such as telecommunications, that are particularly capital and research intensive.

SUPPLY-SIDE AND DEMAND-SIDE SUBSIDIES

Government subsidies can be applied on the "wholesale" side of the market to expand supply and lower private costs or on the "retail" side to stimulate demand and reduce prices to users.

A supply-side strategy requires the government to pick the probable winners when the market is young or to risk extended delays through administrative and legal processes. Market-like procedures, such as auctions, can help direct resources to high-value suppliers and hasten the transfer. Regulatory bodies, which have encouraged forms of price discrimination that direct suppliers' revenues to favored users and services, have found that the concentration of divested interests that became entrenched as a result of such policies makes it difficult to terminate "cross-subsidies." The relatively recent shift away from a regime of cost-based telecommunications regulation to a policy based on incentive regulation, combined with the shrinking size of the monopoly sector, is reducing the use of off-budget cross-subsidies.

User-based "retail" subsidies place purchasing power directly in the hands of final consumers, giving them the choice of vendor and technology. An effective demand-side strategy requires network pricing mechanisms, so that user values can be signaled to suppliers. While long established for the telephone network, user pricing of services has yet to develop in many computer networks. Also, a program of end-user subsidies incurs relatively high administrative costs. Funds must be closely targeted to reach intended beneficiaries; otherwise, resources are dissipated in the form of inefficient funding of unintended uses.

A STYLIZED MODEL

We can distinguish the effects of supply-side and demand-side subsidy strategies by considering a highly stylized view of the economic evolution of a telecommunications network. Figure 5 shows both the average cost of a connection and the average value subscribers place on being connected to the network as a function of the percentage of the market coverage. In these stylized terms we can represent the change in cost per subscriber and value per subscriber during the time that the network grows from its inception toward 100 percent penetration.

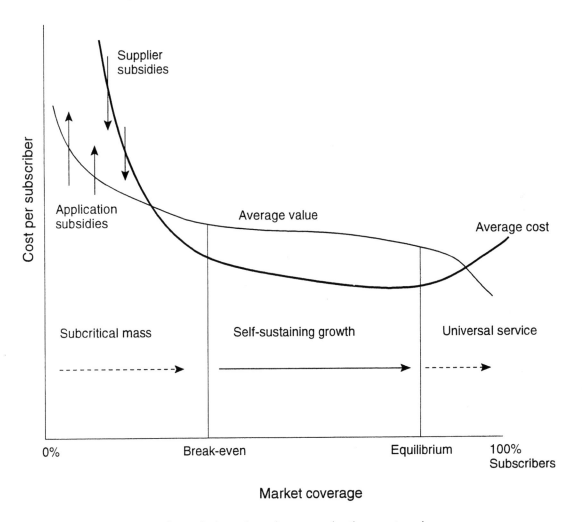

FIGURE 5 Economic evolution of a telecommunications network.

Initially, the average cost of a connection declines with network size, a cost improvement that combines several effects. As a result of "learning by doing," the network suppliers improve production activities and achieve lower unit costs. At larger scales, higher-capacity technologies begin to substitute for the initial transmission and switching facilities. Because the growth in penetration occurs over time, technological improvements and innovations that reduce average costs first become available when the network is larger. Eventually, when the network is mature, "dis-economies of scale" from switching connections between a larger number of subscribers and the higher costs of extending the network to low-density and distant regions set in and cause average costs to level off or turn up.

On the demand side, the average user's valuation of a network connection tends to decline with network size. The first subscribers are those who highly value communication with a small community of interest. Consumers with somewhat less to gain defer subscribing until the price has dropped. To some extent, network growth itself causes the average value of membership to increase—as more subscribers join the network and connectivity spreads, a connection becomes more valuable to existing subscribers and their network usage expands. (One recent example is the fax subnetwork.) Users, as well as the network supplier, also gain from learning by doing and discover unanticipated uses and value in network communication. The demand side of the market also experiences innovation over time, as novel applications are developed and complementary services take hold. Finally, as market penetration approaches high levels, the remaining nonsubscribers may perceive only limited value in network membership and the average value of a connection declines further.

The social value of the network at different stages of development is indicated by the difference between the average user value and the average cost, scaled up by the number of subscribers. Viewed statically, the network is unprofitable to suppliers of network services at low penetration rates. Assuming a common price is charged to all subscribers, profits are negative in the region to the left of a break-even point such as that shown in Figure 5. If, using nonlinear pricing, suppliers are able to capture a higher fraction of the average value per subscriber, they will begin to earn profits at penetration rates somewhat before the break-even point is reached. Beyond this point suppliers will strive to expand the network, potentially competing up to an equilibrium penetration point at which profits fall to zero.

Both supply-side and demand-side subsidy policies can encourage network growth. In a young network, supply policies lower the average cost of a subscriber connection. Demand-side subsidies reduce the consumer's price of subscribing or support applications that increase the value of subscription and thus shift the value curve upward. Each intervention reduces the break-even size of the network and helps get the industry to a point of self-sustaining growth at an earlier date.

In the mature phase of network development, subsidies can assist in pushing penetration toward achieving universal service. User subsidies, particularly if targeted to nonsubscribers, reduce the net price of subscription and effectively shift the market demand upward.

The case for public subsidies to the telecommunications infrastructure turns on the extent to which private markets fail to achieve efficient outcomes and the policy objective of ensuring widely available service. Subsidies to both suppliers and end users can, in principle, improve on market outcomes. But such interventions also result in administrative mechanisms and political interests that increase costs and dilute effectiveness. As a result federal subsidies are purchased with high-cost dollars, and the benefits of a federal subsidy policy need to be substantial and clearly established if they are to compete successfully with other demands for public funds.

Telecommunications Infrastructure from the Carrier's Point of View

Charles L. Jackson

Here I discuss the carrier's point of view regarding public investment in the telecommunications infrastructure. Although I am not associated with any particular carrier, I have worked closely with many different carriers over many years. As a result, I think I provide a unique point of view, based partly on interviews with carriers and partly on my own normative analysis. I see three very important points regarding telecommunications investment. These points are vital to effectively planning for tomorrow's telecommunications infrastructure.

POINT ONE: PRIVATE-SECTOR INVESTMENT IS THE KEY TO TOMORROW'S INFRASTRUCTURE

It is critical to realize that the private sector will have by far the greatest influence in developing the telecommunications infrastructure. Historically, government investment in telecommunications has largely been limited to military and research applications. More recently, growing public concern about U.S. economic competitiveness and social services has yielded several initiatives for federal government investment in the telecommunications infrastructure. The Clinton administration is promoting the development of a national information infrastructure to connect end users to vast amounts of information. However, the administration correctly acknowledges government's role as secondary to that of the private sector—one that serves to "complement and enhance" private-sector efforts. Although such secondary federal support does aid infrastructure development, it can have nowhere near the impact of the private sector. Any federal government investment in the telecommunications infrastructure will be dwarfed by private-sector investment. To illustrate this point, Table 4 contains estimates of annual revenues and investment by major players in the private sector in recent years.

These amounts far outweigh the public investment that can be provided through direct government investment programs, such as the High Performance Computing and Communications Initiative (HPCCI). My (conservative) estimate of private investment [circa 1991] is over $45 *billion*. In contrast, the annual budget for the National Research and Education Network (NREN) program, which is the HPCCI component that most directly advances the telecommunications infrastructure, is about $123 *million*.[1] There is some call for federal subsidies for telecommunications users as a way to expand the infrastructure. Proposals to expand NREN would

---

NOTE: The contributing author is Kirsten M. Pehrsson of Strategic Policy Research.

TABLE 4 Private-Sector Revenues Compared to Capital Investment

	Annual Revenues ($B)[a]	Annual Investment ($B)[b]
AT&T	34.4	2.4
MCI	8.3	1.4
Sprint	5.4	1.2
Local exchange carriers	82.2[c]	19.9
Cable industry	19.8	2.3
Cellular industry	3.1	2.5[d]
CIM[e]	n/a	15.4[f]
TOTAL	>150.0	45.1

[a]Sources: AT&T, MCI, and Sprint: FCC (1991); local exchange carriers: USTA (1988); cable industry: Paul Kagan Associates, Inc. (1992); cellular industry: Leibowitz (1992); commercial, industrial, and military communications equipment: EIA (1991). Figures reflect 1991 except where otherwise noted.
[b]Sources: Same as sources for revenues, except MCI and Sprint figures from Standard & Poor's (1992). Figures reflect 1991 except where otherwise noted.
[c]1987 figure—total operating revenues.
[d]Derived by subtracting total capital investment as of December 1990 from that of December 1991.
[e]Commercial, industrial, and military communications equipment.
[f]1990 figure—estimated by Electronic Industries Association Marketing Services Department.

support bringing schools, libraries, and other new users on to the network and training new users once they are connected. However, there has never been any comparable funding to support telecommunications users in the past. And even if such proposals were to be adopted, federal subsidies of telecommunications usage would likely be so small as to do little to promote the infrastructure. A direct usage subsidy of $5 billion per year would affect only about 3 percent of our nation's spending on telecommunications.

The dramatic contrast between potential public and private investment highlights the importance of creating an environment that invites private investment. This brings me to my second point.

POINT TWO: REGULATORY POLICY IS CRITICAL TO PRIVATE-SECTOR INVESTMENT

Regulatory policy is an extremely important factor in creating an environment that favors investment in the infrastructure. In fact, any major aspect of regulatory policy is likely to have a much greater effect on the nation's telecommunications infrastructure over the next decade than could any politically feasible level of government investment.

There are many examples of regulation's enormous influence over telecommunications investment. The Federal Communications Commission's (FCC) recent rules for personal communications services, for one, are bound to have a tremendous impact on the evolution of wireless communications for decades to come. Limits on market entry (e.g., the cable/telephone company cross-ownership bar, local exchange carrier franchise requirements) restrict areas in which certain telecommunications players may invest. And even within the area to which investment is restricted, regulation may pose other deterrents to investing. Prudence reviews threaten a "heads-you-win, tails-I-lose" situation for the carriers, whereby profits on successful risky investments are limited, while at the same time losses from unsuccessful risky investments are disallowed from the rate base. Depreciation schedules that lag behind actual economic depreciation also deter investments.

Any one of the many facets of regulation may have an enormous impact on incentives for investment in the infrastructure. Consider, for example, depreciation policy. The amount of depreciation currently allowed could be adjusted to more closely match the decline in economic value of the resource being depreciated. As a result, investors would be more likely to get their money back from a resource before it becomes obsolete. This would serve as a catalyst to investment. And depreciation adjustment would encourage additional investment not merely by the amount imposed in additional consumer charges each year, but by several times that amount. The dollars affected by even a slight change in one aspect of regulation—such as the acceleration of telephone company central office depreciation from, say, 18 to 15 years—is comparable to NREN's total annual funding.

Due to the tremendous effects of regulation on investment, it is important to look ahead and consider how regulation is likely to affect carrier investment in the future. There are several possible scenarios that would deter investment. Carriers could find themselves in a world where they are forced to unbundle services. While facing massive competition, they may not be allowed to take measures to confront that competition. They could be in an environment that prevents them from recovering investments. In that case it might be rational for carriers to enter a "harvest" mode. They would reap whatever income is possible from current capital investments but not make any additional investment. As a result, the telecommunications infrastructure would begin to wither. It would then be left up to new entrants in the telecommunications market to build the network for the twenty-first century. No carriers have yet adopted this strategy (at least not openly). But given the current direction of regulation and competitive entry, this scenario is not unimaginable.

POINT THREE: THE PUBLIC SECTOR HAS A ROLE IN INFRASTRUCTURE DEVELOPMENT

The fact that regulation will have far greater impact on infrastructure investment than will public sector infrastructure investment does not mean that certain types of public-sector support are not important—even critical. Government initiatives in telecommunications research and development and promoting applications are beneficial. History shows that government support of research and development has generated important telecommunications technology developments. Over 20 years ago the Advanced Research Projects Agency (ARPA) of the U.S. Department of Defense built the ARPANET. ARPANET's packet-switching technology provided the foundation for today's Internet, which provides nationwide public access to an international network. Federal funding supports the Massachusetts Institute of Technology's Lincoln Laboratory, which has pioneered important developments in telecommunications electronics, including communications satellite technology. Although benefits from investments such as these are difficult to quantify, they advanced the state of the art of telecommunications. Past government funding of facilities has also promoted the telecommunications infrastructure. The Rural Electrification Administration's lending program helped extend telephone service to remote rural areas. Federal funding helped establish the Alaska telephone system. The National Science Foundation's effort to network research centers to supercomputers has provided the backbone to the Internet system.

Government funding in the future will be important because market failures happen. There will continue to be areas of telecommunications where the private sector lacks the incentive to make investments that would benefit society. Government spending can make a significant contribution in those areas.

The Clinton administration is already targeting many of the areas where public-sector support will be beneficial. For example:

- *Research and development.* Public support of research and development of fundamental technologies will help to advance telecommunications state of the art. Recall the extent to which ARPA funding appears to have driven computer and network technology. As of late 1993, funding for the ARPA-led Technology Reinvestment program had fostered numerous proposals for information infrastructure technology development as well as applications.

- *Applications.* Another important role for government is in applications development, which serves to bridge the gap between new telecommunications tools and the user. For example, standardization of electronic medical and other types of forms would help promote use of telecommunications services.

- *Trials.* The government can play a unique role in implementing large-scale trials of new applications. Past successes include the ARPA/Internet and electronic mail.

- *Electronic publishing/access.* Government can make more information electronically available to more users. It can do this by serving as publisher and central server of public information. For example, the government could publish statistical abstracts and provide them via a central server or CD-ROM. Or it could provide access to FCC rules and filings on a central server. Currently, the Office of Management and Budget has a new policy to encourage agencies to increase citizen access to public information. And the White House has become accessible to the public via electronic mail.

- *International parity.* The federal government should promote international parity of telecommunications markets. U.S. telecommunications suppliers operate in many countries. A telecommunications monopoly in one of those countries might use its position as service provider to influence consumers against buying from a U.S. supplier. Government policy-makers can be alert to situations where market distortions may disadvantage U.S. suppliers and may help to neutralize such a situation.

CONCLUSION

Public investment in telecommunications infrastructure is unlikely to be of sufficient scale to make a significant difference in the development of our nation. I have discussed areas in which the federal government can contribute to infrastructure development. However, there are other areas that are obviously not candidates for public investment. For example, intercity fiber networks are not a candidate—the private sector has already built five or six fiber networks.[2] Local fiber distribution networks do not appear to be a candidate for federal investment because they are too expensive for feasible additions to the federal budget. These types of investments in the infrastructure must come from the private sector. A look at history, federal budget realities, and the existing infrastructure indicates that government investment in the telecommunications infrastructure per se is likely to have little impact. However, government initiatives that expand our ability to *use* the telecommunications infrastructure promise significant benefits. Although carriers may engage in some such research, their incentive to do so is limited by how much they think they can capture the benefits of that research.

Changes in regulatory policy are the only realistic means of increasing infrastructure investment by any effective amount and, as I discussed earlier, may be necessary to maintain the status quo. Therefore, it is critical to both preserving and advancing the nation's infrastructure that regulatory changes encourage private-sector investment.

REFERENCES

Electronic Industries Association (EIA). 1991. *Electronic Market Data Book.* Electronic Industries Association, Washington, D.C.

Federal Communications Commission (FCC). 1991. *Statistics of Communications Common Carriers.* U.S. Government Printing Office, Washington, D.C.

Federal Coordinating Council for Science, Engineering, and Technology (FCCSET), Office of Science and Technology Policy. 1992. *Grand Challenges 1993: High Performance Computing and Communications, The FY 1993 U.S. Research and Development Program.* Office of Science and Technology Policy, Washington, D.C.

Kraushaar, Jonathan. 1993. *Fiber Deployment Update—End of Year 1992.* Industry Analysis Division, Common Carrier Bureau, Federal Communications Commission, Washington, D.C., April 30.

Liebowitz, Dennis. 1992. *The Cellular Communications Industry.* Donaldson, Lufkin, Jenrette, New York.

Paul Kagan Associates. 1992. *Cable TV Financial Databook.* Paul Kagan Associates, Carmel, Calif.

Standard and Poor's Corporation. 1992. *Corporation Descriptions.* Standard and Poor's Corporation, New York.

United States Telephone Association (USTA). 1988. *Statistics of the Telephone Industry.* United States Telephone Association, Washington, D.C.

NOTES

1. Estimates for 1993 from FCCSET (1992), Figure 6, p. 28.
2. Major fiber networks include those of AT&T, MCI, Sprint, and the local exchange carriers. The FCC reports that interexchange carriers other than AT&T, MCI, and Sprint have more than 15,000 route-miles of fiber. See Kraushaar (1993).

An Information Provider's Perspective on Government Investment in the Telecommunications Infrastructure

William F. Gillis

I have been asked to comment on the benefits, if any, of government investment in the so-called information highway. I will do so from the perspective of an information provider who soon will be introducing some of the services that can utilize this much-discussed facility. This perspective is shaped by my experience in many of the business sectors upon which this workshop's discussions will have direct impact—consumer electronics, pre- and post-divestiture telecommunications, software, and on-line services. I operate on the basis of the marketplace being the final arbiter of discussions like those in which we workshop participants are engaged. I am a supporter of a free, open, and competitive marketplace unencumbered by historical beliefs and restrictions.

Here I address several questions that have been put to me:

- Should there be any (or more) government investment in infrastructure, facilities, or services?
- Is there a benefit from government-provided seed funding?
- Do government-supported services really put a meaningful damper on investment in new or better services?
- How much government investment would be too much?

Before directly addressing the questions at hand, I offer the observation that we cannot and should not define the ultimate delivery mechanism in terms that are too restrictive. To speak of it only in terms of copper or fiber presupposes that the marketplace is unaccepting of additional delivery methods offered by various wireless and broadcast techniques.

SHOULD THERE BE ANY (OR MORE) GOVERNMENT INVESTMENT IN INFRASTRUCTURE, FACILITIES, OR SERVICES?

I firmly believe that government investment should be directed to areas in which it serves to support a free and competitive market. This would include investments in the timely creation of standards and to ensure network interoperability. It could also include the establishment of testbed facilities but should not include the creation of services.

I have not observed conditions that traditionally suggest that government investment is warranted in infrastructure per se. There have not been either marketplace failures or a demonstrated lack of ability or of willingness on the part of the private sector to invest.

204

IS THERE A BENEFIT FROM GOVERNMENT-PROVIDED SEED FUNDING?

For the government to begin investing in the creation of commercial services is not necessary. There are ample services being investigated and developed in the private sector. There is, however, a potential role for government to provide seed funding to encourage the deployment of variations of selected services—in the educational setting, for example.

By way of illustration, two years ago while heading an unregulated unit of a telco, my unit conceived and developed a fax machine-based social studies learning aid. It was deployed in two inner-city schools in Denver. It was an immediate hit with students and faculty alike. Not only was interest in the curriculum heightened but there was also a dramatic reduction in absenteeism and a measurable improvement in writing skills attributed to the project by teachers overseeing it. As a result of these positive results, there was a strong interest in deploying the course throughout the city's school system. Unfortunately, an opportunity was lost—no government funds were available to expand the program to other schools despite the equipment providers', curriculum designers', and our willingness to provide our services on a greatly discounted basis. If ever there was an opportunity that warranted government investment, this was one. It sits on a shelf, its promise unfulfilled due to a lack of funding. This is an area in which seed funding would have been both productive and appropriate.

Seed funding in the area of acquiring equipment for classroom use and the training of instructors in its use is also an appropriate investment for the government to make. Consider, many schools across the country have recently invested in computer equipment and have begun integrating it into classrooms. Much of it will be rendered obsolete by some of the services contemplated as the basis of this workshop discussion. Much of the equipment in place in schools today is not suitable to display full-motion video or generate the stereo sound of services.

DO GOVERNMENT-SUPPORTED SERVICES REALLY PUT A MEANINGFUL DAMPER ON INVESTMENT IN NEW OR BETTER SERVICES?

Although there is no universal answer to the question of whether government-supported services put a meaningful damper on investment in new or better services, there is a distinct risk that government-supported services could limit the introduction of new and improved services. One of the primary incentives to the creation of new services is that they, due to the diligence of their creators, offer a degree of uniqueness in the marketplace that, in turn, offers the hope of substantial return on investments. Were there an existing or contemplated government-supported service that offered competition to a proposed new service, would venture capitalists and corporate investment organizations be as willing to invest as they are today? I believe not.

Not only would this present a reduced earnings scenario, but it could also have other impacts of significance. Among these are:

- Fewer patent disclosures and patents granted;
- Slower and more widely spaced introductions of newer generations of products/services; and
- Fewer creative ventures and small businesses formed.

On the latter point, I believe that any action that could forestall the creation of new small businesses should be avoided. Small businesses continue to be a strong source of job creation in today's service-based economy.

In lieu of direct government spending on applications and services, rethinking investment tax credits and incentives may be a more beneficial form of government investment.

There is an alternative area in which government investment would be beneficial—reorganizing its vast databases of information to make them more useful to the public. I read with interest the proposed National Information Infrastructure Act of 1993 and its stated intent to digitize government data to allow public access. I applaud this effort but also encourage us not to stop at simply digitizing the data. More importantly, I believe it is necessary to reorganize the data into a uniform structure that allows the data to be more efficiently accessed and searched. Successful completion of this effort would simultaneously reduce access charges to constituents who use the services and reduce the government's future data input and printing costs.

Once the data are suitably reorganized, two other areas of investment would seem appropriate. First, investment should be made in developing descriptive terminology and nomenclature that is readily understood by nonspecialists. Refer to a ten-penny nail as just that and not with the "fastener-interfibrous galvanized" terminology that I encountered during a recent search. Next, once the nomenclature is revised into everyday English, on-line directories should be made available to facilitate searches. I recognize that this is a mammoth undertaking, but it is both necessary and consistent with the aims of legislation recently contemplated and becomes only more complex and unwieldy the longer we wait.

HOW MUCH GOVERNMENT INVESTMENT WOULD BE TOO MUCH?

Obviously, there can be no definitive answer to the question of how much government investment would be too much. In general terms, government investment should be sufficient to support creating interoperability standards, revamping its vast storehouse of data, and providing classroom facilities for utilizing the newly available information. These benefits and, I dare say, market-driven aspects of such possible legislation as the proposed National Information Infrastructure Act of 1993, are most appealing.

With the government concentrating on its role as a facilitator, those of us who are charged with extracting value from visions can set about the task of delivering the products and services that the market demands—products and services that save time and money and allow Americans to better manage their lives on both personal and professional bases.

Finally, unlike a description I read recently that indicated that legislation such as the proposed National Information Infrastructure Act of 1993 could help unleash an information revolution, I submit that it holds the promise of harnessing runaway data and allowing it to be delivered as easily accessible and useful information.

Economic Dividends of Government Investment in Research and Technology Development

Robert E. Kahn

In addressing the role of government with respect to information infrastructure, I start with the assumption that a role for government not only exists but also is essential; but there is no formula for it yet.

Second, the national information infrastructure (NII), as we have all been referring to it, is more than just the networks and the computers that attach to it; it includes the higher levels of this infrastructure and the applications that sit on top of it. I tend to describe the NII in terms of the core, which contains the generic portions of infrastructure; the mantle, which contains the domain-specific portion; and the outer shell, which contains all the systems and applications that run on top of it. I also tend to describe the elements that are at the network level as being the lower level. Many people think that the network is all that there is in the infrastructure. Identifying it as the bottom layer implies that there are higher levels, too, and that there are applications that you can then put on top of that.[1]

But if all you have is a network on which to build end-user applications, it is no different than taking a personal computer (PC) as a plain piece of hardware (without systems software) and writing an application on it. If somebody else were to write an application on it, there would be no guarantee that the two applications would work together, because each person probably had to deal separately with all of the system problems for their application on this raw hardware. Each would have or develop separate naming conventions, file systems, buffer management, and all of that. So if you put the two programs in the same machine, they would surely collide with each other.

At least a few companies have made a big business out of building what I call the "middleware" of the personal computer industry. I think there is a similar problem and opportunity in the NII to build the middleware—namely, the generic and applications-specific services.

I also see the NII as evolving more like the economy than a building to be architected. I don't think there will be an architectural blueprint for the NII as there is for a building such that, having produced the blueprint, you then have somebody construct the NII. Instead, I believe we are going to architect a framework in which the NII can evolve as a technical development; this will allow market forces to develop on their own. The framework will probably involve notions of standard interfaces or standard objects and the like. But fundamentally the NII will behave like an economic system. It is going to have a certain degree of Brownian motion associated with it, but hopefully it can be guided by user needs and requirements.

Most important is going to be the process by which the NII is managed and evolves. A critical part of that, from the government's perspective, is the oversight that it can provide in ensuring that the process maintains a level playing field, doesn't disadvantage parties, and so forth.

My comments focus on the research and development (R&D) side of the picture. And although I was pleased to hear Walter Baer mention some of the work that I had done on the use of vouchers to take technology from a maturing state into commercialization, I am going to focus now on how we take a technological idea and get it to the point where the technology reaches critical mass. If there is one example that embodies all of the attributes I want to touch upon, it is the Internet.

The genesis of this concept may be traced back to an idea that existed within the Advanced Research Project Agency (ARPA) back in the early 1970s. I was fortunate enough to be involved in that. We watched a network of networks grow from a concept to something that is probably going to have more than 100 million users on it by the end of this decade. It is a real success story in many different dimensions. The amount of government money that has gone into nurturing it is a tiny fraction of the total sum that is now spent on it.

Now, you probably know that some of the carriers at first saw the Internet as their competition. It took a while before they realized that the Internet was really their market. So the question, then, was how to co-opt it. But in fact the Internet is likely to be part of the larger NII, and what they really want is to play a major role in providing the NII.

There are multiple competing forces as we develop this infrastructure. One force tends toward early commercialization. Once a capability has reached a critical mass, it really wants to grow. In the case of the Internet today and what is happening on it, there are forces to spread connectivity to more places, to get it to the schools, to get it to the small businesses. This is the kind of demand for which low-bandwidth lines or even an integrated services digital network (ISDN) service would suffice, especially if the users were only going to use e-mail. For many who don't use networks today, even that would be a major step forward. Another set of people, those who are dealing with the most advanced supercomputers and moving billions of bits a second on gigabit nets, are looking forward to going to teraflop machines and terabytes per second.

So there is a market pull on the low-end side as the usage base expands, but also a technology push at the top to take into account new technology opportunities or maybe even high-end application opportunities. Both of these are important because this is where the market is, and this is where the future is, in some real sense.

R&D AND COMPUTER TECHNOLOGY

Let us focus on how R&D can move the cutting edge forward. In addition to creating new capabilities, R&D can help to expand the market and improve services, making things more reliable and integrating commercial offerings with advanced research capabilities. These are essential elements of what government research funding can enable.

My focus here is R&D rather than long-term infrastructure development in the sense of building networks and maintaining them. But long-range R&D on computer architecture and, in fact, anything that is related to NII technology should be coupled to prototyping key architectural concepts. If you have an interesting concept that people have developed or on which some basic research has been done, you may want to prototype that concept, build pieces of it, a chip here, a cache there—whatever it is, try it out, see if it works. You may want to actually build a whole experimental system. If a systems idea is worthwhile enough to evolve but is not likely to be generated spontaneously in the marketplace, building an experimental version is a useful way to proceed. If you build a suitable testbed, you would like to get it used in small numbers by a few groups of people who are knowledgeable in the area and also real users.

Those are all viable and important roles for the government to play.

Now, there are some unique mission-oriented requirements where the government literally has to do more than that (e.g., concerning space and embedded computing for the military). But in general it seems to me that the government ought to avoid funding production development of computing. They ought to piggyback on commercial development whenever possible. A couple of computer-related examples can put this into context.

Case A involves computers that have been developed with direct government support. The ILLIAC-IV project was an ARPA effort that started in 1965 and produced a working parallel processing machine that was built with government support. Although many people thought it wasn't a success because there was only one of the machines (later dismantled), the project made several crucial contributions. It demonstrated to the world, at a time when a million instructions per second (MIPS) seemed like a very large amount of computation, that it was possible to build a machine at the hundred-MIPS level. ILLIAC-IV was actually used for real work at a time when you couldn't buy any commercial supercomputers, because no companies were making machines at that level. It also stimulated the commercial production of semiconducting memory.

The Connection Machine was another example where ARPA created a parallel processing computer that had 64,000 processors on the way to a million processors. Again, this was an attempt to actually develop a machine. Yet in late 1993 it was unclear what the long-term commercial viability would be of the Connection Machine, but Thinking Machines Corporation [which filed for bankruptcy in 1995] was then trying to sell those machines in the commercial marketplace. The government also invested in the Butterfly machine at Bolt, Beranek, and Newman. This is not a machine that succeeded in the commercial marketplace.

In the three examples above of government involvement in funding production development, the results are mixed. In some cases it really is important for the government to be involved; in these, not.[2]

Case B, which involves indirect government support, includes areas where the involvement of the government is crucial, but the government never set out to fund the development of these machines. For example, some of the Cray machines have been developed basically in close liaison with government-supported national laboratories. But instead of setting out to fund the development of a new supercomputer, the government involved key researchers at the national laboratories to generate concepts for a new machine and make specifications available to industry. The government implicitly or explicitly agreed to buy a few of them when they became available, if some company (e.g., Cray) would build them. Their initial production, in fact, started a process in motion to make them available commercially to a larger set of customers. The government didn't contract for the development of the machines, but it did agree to buy a few of them if the machines were produced and met the specifications.

Sun Microsystems and Silicon Graphics are two spinoffs of ARPA support. Sun Microsystems, a company that is now the largest producer of workstations in the United States, got its start as a small ARPA-supported research project at Stanford. (The Sun acronym stands for Stanford University Network.) ARPA had funded a small project at Stanford to do innovative very large scale integrated (VLSI) circuit architecture work, and the principal investigator, Forest Baskett, needed some VLSI design workstations. After having looked at what was likely to be available from other sources—research and otherwise—they decided they needed to develop their own capability and had a good idea for how to do it. A student named Andy Bechtelsheim had an interesting design for a frame buffer system. Bill Joy at Berkeley had been funded by ARPA to build a virtual memory version of UNIX. ARPA gave Stanford a very small amount of money to prototype a tool, basically, rather than a computer, for doing VLSI design. That work became the basis for Sun Microsystems.

Silicon Graphics actually had its genesis in that same effort. Jim Clark, who until recently was chairman of the board of Silicon Graphics, designed a chip that could manipulate graphic images. He called this the geometry engine, and based on that he then set up a company that could manufacture low-cost graphics workstations with the geometry engine as an integral component. Silicon Graphics is now a leading producer of high-quality, low-cost graphics workstations.

Those are some examples where government involvement was critical, but the government didn't set out to actually build machines and in most of those cases not even to buy them, as in the Sun Microsystems and Silicon Graphics examples, although it did so with Cray.

Case C includes examples where there has been no government support. One example is the personal computer, which didn't start from the government building the first personal computer. The microprocessor is a second example. The interesting thing about these two developments is that they took off by themselves because there was a latent market for something at that level, given that it could be built.

R&D AND NETWORKING

There are three stages to the development of networking. The first stage is what I call the seed stage, in which you have an idea and are trying to get it explained, just like with the machines mentioned above. The elements of that stage are technology development (generally experimental technology development, because you are building things that need to work), testing and evaluation (generally through hands-on experience, initially in the laboratory), and field testing deployment, hopefully with some friendly real users, and pilot projects.

If, as a result of this seed effort, you learn that there are real users and that the technology really works, you go into a second stage, an expansive mode, where you get more users, conduct further tests, and refine the technology, give it more capabilities, make it more stable, make it more robust—scale it up and iterate. What is happening at this point is that a small market is being nurtured. One of the critical questions later is how to wean that market from something that may be provided as a free good. This often occurs when the government has been involved in setting a network up and operating it or funding it for its contractors.

The third stage is to commercialize the technology, given that it has real commercial potential: it has been demonstrated, it has a reasonably sized user base, it provides a meaningful capability, and one can see how it can expand more broadly. Hopefully, commercialization can be done without the government actually having to take a direct role in leading the effort.

Now, let us consider the critical mass phenomenon. There is a period of time before the infrastructure is actually deployed, when R&D on it is being conducted. When the technology infrastructure is ready, major R&D money has already been invested. More funding is needed for deployment of resources and for staffing the operation. As the user base starts to grow, additional R&D money is needed for developing applications to run on it, for testing and refining the technology, and, at some point, if the market continues to develop, critical mass is reached. And then, around the critical mass point, the risk flow turns positive, and one can start to make a profit, although it may take a while longer to break even. So there is a big up-front cost in deploying infrastructure, which could be as long as 20 years in some cases. The private sector won't invest until the technology is close to the positive side of profitability.

Now, there are different kinds of government involvement in networking infrastructure. One kind is direct hands-on involvement—I call this procurement. The ARPANET embodied this kind of model. No computer networks existed, and nobody was taking steps to develop them (apart from continued use of dial-up telephone lines). The government just jumped in—and it could afford to do so, a crucial element. It leased 50-Kbps lines from AT&T. The government funded Bolt, Beranek, and Newman (BBN) to develop switches, because there were none in existence—these were the very first packet switches. They then procured the switches from BBN. The government also took responsibility for managing the network, for controlling access to it, and for developing ARPANET policy. The day-to-day management was contracted out, but the government remained in charge. This is an approach I would characterize as a procurement network.

Project involvement is a second kind of approach to infrastructure deployment. This approach is more characteristic of the early stages of the Internet, which began with the ARPANET as a central core but evolved with research networks, gateways, and local area networks connected to it, mainly for research initially. During the initial phase, almost all of the R&D was funded by the federal government. The initial networks were government-procured networks, but soon thereafter private-sector networks were connected as well. The gateway was provided by both parties. The government became very much involved in managing the process by which the Internet evolved. It controlled the handing out of addresses and the standards process, and so actually had a tight rein on the Internet system. There was no commercial use during this early phase.

Another kind of approach is benevolent partnering. In this example the government's role is not the dominant one. Rather, the government manages selected elements of the process, rather than the whole process. I think this approach characterizes the NSFNET and certainly characterizes the Internet development. In benevolent partnering, the R&D is funded by both the private sector and the federal government. And, in fact, the standards process is starting to involve private-sector efforts as well as government-related efforts.

The final approach is one that I think is going to be more applicable to the national information infrastructure. I call it oversight and steering. In this approach there is both private sector and government-funded R&D. The standards development is carried out largely in the private sector, perhaps overseen by the federal government in some sense. There is widespread commercial use. Instead of the government managing selected elements of the process, it now provides oversight, because the actual management of the enterprise can reside in the private sector as well.

We have some experience with this approach in the context of the Internet. When the Internet started, it was an R&D project within ARPA. For the first 10 years or so, the project was run by just one or two of us within ARPA (Kahn and Cerf). Few people had yet discovered the power or the latent potential in this technology.

Around 1978, as the technology deployment was continuing, I got worried that if Vint Cerf, who was the program manager and who had most of the project details in his head, got hit by a truck, we would have a real problem. I thought we needed to get the community involved in a more direct fashion. And so Vint formed something called the Internet Configuration Control Board, or ICCB. The ICCB helped in many ways but mainly was in the path of the information flow. It could be aware of developments while they were in planning and even help to plan them.

In 1983, shortly after the split of the ARPANET into two roughly equal parts—the MILNET and a residual ARPANET—it became clear that the ICCB meetings were too big to allow the ICCB to serve its original function. Instead of 12 ICCB members meeting around a table, an order-of-magnitude more people were coming to listen to the deliberations, because they had become interested in networking. Coordination of ICCB meetings had become unwieldy. I had briefly retained Vint Cerf as program manager for the Internet activities in late 1982 and handed the responsibility to Barry Leiner about a year later.

One of Barry Leiner's first acts was to set up something called the Internet Activities Board, or IAB. In this new model the work that had been a subject of discussion in the ICCB meetings was farmed out to task forces. There were 10 task forces originally. They reported to the IAB. That process actually worked very well. Today there are close to 100 different task forces involved in the continued evolution of the Internet. It was no longer possible for the IAB to coordinate this growing set of activities. And so something called the Internet Engineering Task Force (IETF), which was a task force of task forces, got created with Phil Cross as its chair.

In the 1986 time frame, when this was becoming the mode of operation, there was no government body directly involved. In 1986 ARPA decided to get out of networking. ARPA literally got out of it for a very short period, and, during this period, NSF took steps to lead the federal networking activities. So there was a period in which no part of the government was responsible for this effort; it free-floated.

During the 1987 to 1988 time period, the U.S. government became involved once again. A federal network was established (originally called the FRICC and later the Federal Networking Council). DOE, NASA, ARPA, and NSF funded a secretariat, whose job it was to make sure that these Internet-related processes kept going and that the government maintained its involvement with the process. The IAB ended up with not only an engineering task force, the IETF, but a small research task force as well.

Today, this process has evolved still further. The U.S. government funds an IETF secretariat and separated other parts of the Internet standards process. It continues to administer addresses, and it supports several network information services. However, the process itself has gotten more mature. And, along with its educational and scientific uses, the commercialization of international usage of the Internet has grown rapidly. The Internet Society (ISOC) was recently set up as a professional organization to take responsibility for adopting and promulgating the standards, for official publication of relevant information about the Internet, and for holding an annual meeting to bring together the relevant parties. In addition, ISOC engages in a variety of nonprofit activities that help to promote the Internet in a professional society context.

The old IAB was replaced by a new body with the same acronym, the Internet Architecture Board, which is part of the Internet Society and serves as an appeals court for the standards process. Adoption of standards is taken up by an Internet Engineering Steering Group, which basically is the leadership of the IETF.

As of late 1993, connections between ISOC and the IETF were not formally constituted. This is interesting for a standards-making process, because the IETF meetings are open to anybody who wants to come. You don't have to be a member of ISOC to show up. While the analogy is not perfect, I could liken it to a gathering on the Mall, where one group handles the logistics, another group plans and leads what happens, and another group reports on the events. You don't have to become a formal member of any organization to be present. You just have to show up.

Let me now discuss the chicken-egg syndrome in some of these network R&D activities. The single most important concern, to me, is that if you don't have the technology in place to use the network and if you also don't have the network in place to make use of the technology, how do you get started since you need both of them?

In the case of very high speed networks, the computer industry won't build the machines that send and receive data at a terabit per second or a petabit per second and the carriers won't deploy networks with these end-user speeds until a market develops, resulting in a kind of a deadlock. One way to get started is through the use of testbeds. The Corporation for National Research Initiatives has effectively exploited that notion with support from NSF and ARPA under contributions from industry (and particularly the carriers). Several gigabit testbeds have been created around the country. Many of the Bell regionals have participated and have been very helpful. So have some of the long-distance carriers like AT&T and MCI. Bellcore and IBM have been two of the most active participants. Bellcore has contributed advanced technology including an experimental ATM switch. And IBM also has contributed technology. GTE and Bell are active participants in a testbed involving medical networking applications. Many more of the computer companies were willing to participate both because the government was backing the effort and because the overall direction of the project was right. The federal government put up about $20 million for research to be carried out over about a five-year period. Private industry, it has been estimated, put up something like several hundred million dollars in equivalent costs of their people, their facilities, and their technology over that same period. I believe they would never have done that if the federal government hadn't gotten involved. Industry was able to capitalize on research in the universities, which is nominally where all the government-provided funding ultimately went, by providing facilities.

I would like to emphasize how hard it is to make this kind of collective effort work. Even though a few million dollars for facilities is a small fraction of a large company's budget, it comes out of somebody's budget. Deciding to take that money and put it into an R&D project, which may not have any potential in the very short term, is not something that companies are usually likely to do.

But when the government says that an investment is on the path to a national network, to our NII, then it may become quite a bit more important for them. What is in it for them to invest? Well, they will learn about technology and its applications, and they will get some early experience with deployment of the technology; this is relevant because they think there is a follow-on. They may get to do some interoperability testing much earlier than would otherwise be possible or in contexts not easily achievable. SONET was made to work as a result of some of these testbed activities. Finally, there is the stimulation of applications as well as general business development on their part.

The same kind of thinking process applies to the computer companies. Their motivation in participating in advanced testbeds is to explore advanced technology, develop new knowledge, meet the needs that are perceived in the R&D community, and, in many cases, develop a market for the technology. Certainly all these companies are getting market visibility. Some of them are actually advertising their participation, which may have value in its own right.

Having gone through this process once, I can tell you it is really hard. If we had to do it again, I don't know that we could create another national network initiative and motivate the carriers and the industry participants to do this one more time. We need to find a way to make upgrading the national infrastructure part and parcel of the normal R&D process in this country.

There are things that we can do to help. One element is government funding for some of the early infrastructure. In the case of gigabit testbeds, the infrastructure has all been contributed. No funding was provided for the facilities. I doubt this can be replicated in a larger sense, however. Another element might be tax credits for investment in advanced infrastructure testbeds. Certain rights and imprimaturs that could be granted by the government may also play a role.

Given that these testbeds have gotten to the point where they are viable demonstrations, can they be grown (individually or collectively) to include more of the research community or even to a critical mass state? What are the mechanisms? Is direct support from the government the answer, or is there another way to do it? Are there other approaches that would help us get there just as fast, if not faster and at least as effectively?

REFERENCES

Computer Science and Telecommunications Board (CSTB), National Research Council. 1994. *Realizing the Information Future: The Internet and Beyond.* National Academy Press, Washington, D.C.

Computer Science and Telecommunications Board (CSTB), National Research Council. 1995. *Evolving the High Performance Computing and Communications Initiative to Support the Nation's Information Infrastructure.* National Academy Press, Washington, D.C.

NOTES

1. Editor's note: The conceptual layering of the information infrastructure is described in Chapter 2 of *Realizing the Information Future* (CSTB, 1994). Dr. Kahn was a special advisor to the NRENAISSANCE Committee that developed the report.

2. Editor's note: As discussed in *Evolving the High Performance Computing and Communications Initiative* (CSTB, 1995), commercial success of a given company or a given product design is not a complete or even relevant indicator of the contributions of federal R&D investments. The investments in parallel processing have resulted, as of 1994-1995, in commercial adoption of moderately parallel processing systems and scientific research benefits from the application of moderate and massively parallel systems.

Perspective of the Noncarrier Transport Provider

Laura L. Breeden

As the executive director of FARNET (the Federation of American Research Networks), I work with 35 organizations that are providers of Internet services at the state, regional, and national levels. These organizations vary greatly in size and in their objectives. Some are small nonprofits with only a few employees, focused on serving a single state. Others are large for-profit companies with hundreds of employees and a national reach. In discussing their perspective I will, of necessity, be describing an idealized view that may not fully represent any single one of them. Since a majority of the members of FARNET are small- to medium-sized nonprofit organizations, that is the perspective I will lean toward.

First, some definitions. By "noncarrier transport providers" I mean value-added systems integrators (who do not typically own the transmission facilities and who are not regulated as common carriers). Because I have limited knowledge about them, I have not included cable, wireless, bypass, and similar providers of "raw bits." I define "transport" in this context to mean the movement of digital information from one location to another. Note that "transport" used in this fashion is an increasingly low-margin (commodity) business, although when the market is vast, there is plenty of room for niche providers to prosper. The recent frenzy of mergers and joint ventures among telephone, cable, and entertainment companies attests to the perception in the industry that the most interesting, and potentially most profitable, new developments are in information delivery (of news, games, databases, and so on).

The question before us is, What is the role of federal investment in building the telecommunications infrastructure? For the community that I represent, this question can be approached by asking what supply-side investment has reaped? The history of federal support for the Internet (and its predecessor the ARPANET) demonstrates a movement from direct federal sponsorship of the network (including the research and development (R&D) required to develop it, the equipment needed to connect to it, and the ongoing operational expenses associated with it) toward indirect support (to developers of network services, who then purchase commercial equipment and tariffed circuits, and to network users).

I would argue that, through its policies in the 1980s in this arena, the government stimulated a group of "market makers," to use Richard Mandelbaum's phrase (see Mandelbaum and Mandelbaum, 1992) with their roots in higher education, the computer industry, and research. This group

---

NOTE: At the time she developed this paper, Laura Breeden was executive director of FARNET. Since March 1994 she has been division director for information infrastructure at the National Telecommunications and Information Administration, within the Department of Commerce.

was focused on the needs of the education and research communities initially, and federal support allowed them to develop new services and develop the market for the Internet by reducing risk, guaranteeing a certain level of demand, and creating a critical mass of users.

In late 1993, as we contemplated the development of the national information infrastructure and how the government's investment strategies should change, it became time to ask, "When do you move federal investment from the supply side to the demand side?" (As many other members of this panel have said, it is also important to look at all of the tools at government's disposal, such as tax policy and regulation, since these may be more effective and influential in many situations.)

It is doubtful that if Washington gave a $100 tax rebate to every teacher in America for "networking" (a $100 million investment), the result would be a national research and education network. There is a social organizing principle at work here, which argues for targeted spending aimed at leveraging certain activities within the target community, whether that is K-12 education, libraries, health care, or economic development.

To be effective, federal investment policy will have to be designed to meet the needs of specific user communities, at specific points in time. Generalizations about information infrastructure are becoming rapidly more hollow as new technologies, service providers, and access devices sweep on to the scene. Lessons from the past will be useful only to the extent that they are carefully drawn from particular historical circumstances.

Policymakers will need to practice humility. They will need to be painfully honest about what they know and what they don't. Data, and the recommendations that follow from them, will have to be looked at with a cold eye. A lot of what we heard at this workshop was amazingly vague, imprecise, or not well supported. Further, we need to maintain a sense of scale in policy discussions. The U.S. telecommunications industry spent approximately $200 billion in 1992. Comparatively, as mentioned in this workshop today, defense research, development, testing, and evaluation for intelligence and communications accounted for $5.3 billion, the High Performance Computing and Communications Initiative (HPCCI) accounted for $1 billion, and the National Research and Education Network component of the HPCCI accounted for about $100 million. Typical research university computing cost $50 million, the National Science Foundation's Networking and Communications Research and Infrastructure Division budget was $35 million, and the NSFNET backbone was $12 million.

What can the federal government do effectively? Is there a role for government investment? I believe that government has an obligation to promote equity, in the first place. Further, public support at the right time and in the right place can offer:

- Risk reduction, in high-risk but potentially promising areas;
- Proof of concept (when private capital is not available, because of the risk or because the benefits to be obtained are not likely to attract private investment); and
- Support for network externalities (such as the development of a critical mass of users) and for public goods.

There are a number of ways to achieve these goals, including support for:

- Applications development;
- User connection to the network or information service;
- Development of telecommunications networks and systems;
- Direct R&D; and
- Incentives (e.g., tax policy).

I concur with Walter Baer's excellent test of investment policy as presented earlier at this workshop:

- Involve industry and users in design and deployment;
- Require cost sharing;
- Have an explicit plan for handing off technology to the private sector;
- Focus on meeting technical and economic objectives;
- Shelter from capture by political constituencies; and
- Avoid rigid schedules and time constraints.

I would further recommend that formal evaluation and dissemination be required components of all federal technology and infrastructure investments.

To summarize, I urge that policymakers:

- Examine their data critically and carefully. Data must be accurate, and conclusions that follow must be relevant to specific times and circumstances;
- Admit their assumptions up front, and question them continually; and
- Maintain a sense of scale. Data presented at this workshop demonstrate that some of the areas that have received the most attention from policymakers actually loom small within the overall dimensions of federal activities. Pay attention to the things that matter.

REFERENCE

Mandelbaum, Richard, and Paulette A. Mandelbaum. 1992. "The Strategic Future of the Mid-level Networks," pp. 59-118 in *Building Information Infrastructure*. Harvard University Press, Cambridge, Mass.

Discussion

JOEL ENGEL: I would like to address a couple of things. The first is the whole issue of critical mass, which so many people have spoken about. Surely, there is a critical mass issue. But that doesn't require subsidies. We have many examples, even from very recent history. For example, the very first purchaser of a VCR [video cassette recorder] found it not very useful if there were no cassettes to play on it, and it would not be very good business for people to make cassettes if there weren't very many VCRs. In the same way, without software the utility of PCs [personal computers] would be considerably less than it is today, but it was not very desirable for people to develop software if there wasn't a large mass of PCs and PC owners to sell it to. And yet, those are two examples of exploding industries that didn't require any kind of subsidy to get started, but were started by what we used to call investment.

Wally Baer gave the example that even for the telephone industry, the entire telephone network was built on private investment. The government did invest. It invested in telegraphy, which is the other example of the ability to choose winners.

Bob Lucky gave the figure of about $25 billion as being required to convert the telephone network from copper to fiber cabling. But we ought to put it in perspective. As Chuck Jackson said, just the LECs [local exchange carriers] alone spend about $20 million every year to keep their networks modern. In the last few years about a third of that has been in local distributions, but that is because all of us have been replacing our analog switches with digital switches. That emphasis can shift. In addition, the front page of the business section of yesterday's *USA TODAY* shows that the free cash flow of the LECs ranges from $4 billion to $6.5 billion, which means that even in addition to this shift, if there were good business reasons, those investments could as much as double. So, as Wally Baer pointed out, while we sit in this room and debate whether or not there is a critical mass issue, the private investor is going to inconsiderately knock the legs out from under the debate.

The other point that I want to make with much fervor has to do with the view that one of the necessities is to make the upgraded infrastructure available to all Americans. I don't think that is the critical issue. The critical issue is to make it interesting and attractive to all Americans. I really want to reinforce the point that Bill Gillis made about the importance of applications and content. Earlier today, both Bruce Alberts and George Turin told about having to either hide or lock up their TV sets because they didn't like the content that their children were getting out of it. They didn't say, but from my experience as a parent, I am willing to bet that they didn't have to lock up their encyclopedias. I think that the real challenge is understanding what, when we build the infrastructure, is the content that will make them come.

DAVID MESSERSCHMITT: I think that particularly on the first point—that market mechanisms will take care of the critical mass problem—there may be some comment.

218

CHARLES JACKSON: Just a historical observation. I believe that the first widely known economics paper on critical mass issues in networks, network externalities, is by a colleague of mine named Jeff Rohlfs [formerly] of Bell Laboratories. He looked at the problem by looking at the videophone experience in the late 1960s, when the old Bell system rolled out a product for which they charged $60 a month independent of usage. His analysis was that if the charges had been based on usage so that people had the product on the desk but didn't have to pay for it if they didn't use it, the network might have grown a lot faster.

Economists have tools to analyze these issues. But I think that the insights that we get from network externality theory do give us clues about how to price new services, and maybe some ideas for promotional activities by the government. I think the point in the long run—which complements Bridger Mitchell's point that if you try to subsidize something and then it never quite makes it to the critical mass point, then you still have all these people getting the subsidies—is that you can lock yourself into a very unfortunate and wasteful situation possibly, and we need to be sensitive to that.

But nevertheless, I think that theory can give a lot of useful insights.

MESSERSCHMITT: Would those pricing mechanisms be consistent with making a profit during this critical mass period?

JACKSON: No. What you need at the beginning is a promotional price to get people into the network, and then you profit from it later. And clearly you can imagine industry structures where that second stage isn't possible.

WALTER BAER: I don't think you are hearing any wild enthusiasm from this panel for large government investments in this area. What you are hearing, I think, is that there may be government roles on the R&D side in dealing with some of the network critical mass issues.

The two examples that you gave are interesting. The VCRs and the PCs really are kind of stand-alone products, at least initially. They didn't require networking to be sold initially. And there may be some difference when you get this network externality effect. The other point is that the VCR actually did profit from a fair amount of government-supported R&D 10 or even 20 years before; much of it was, again, defense oriented at the beginning.

LAURA BREEDEN: I agree with the point, though, that the applications that are going to be delivered over the information infrastructure are critical. I heard Larry Irving, who at one point was very involved in the reevaluation of cable before he went to the Department of Commerce, say that the two biggest sellers on pay-for television are tractor pulls and soft pornography. I believe that the government may have an interest in seeing that other kinds of programming are available that perhaps are not driven by the market, because the market either isn't there or is so immature that no single private investor or group of investors can make a profit on those applications.

Having said that, I don't think it is fair for the private sector to say that government should fund all the high-risk application development and that the private sector then will reap the benefit, the profits. One of the things the questioner said was, "let's make it attractive," and that is the hard part. If there is one thing that U.S. industry is good at, it is making stuff attractive. It is creating demand.

So, I think there is an issue of balance here and it is not a black-or-white decision between no government investment and socialized networking.

LUCY RICHARDS: Since I come from an R&D committee on Capitol Hill, I want to make a couple of comments on the issue of government R&D. First, I appreciate the comments on the National Information Infrastructure Act of 1993. That was our committee's bill and is basically a refinement of Public Law 102-194, which funds a lot of the high-performance computing applications. As that bill went through our committee, we tried very hard to refine it, to try to make clear that the government was not trying to compete with the private sector in the development of services or networks. That bill also makes significant changes to the NREN programs, with one of the goals being to make clear that the NREN is not trying to duplicate what private industry can provide.

A couple of other comments on funding. One concerns the issue of manufacturing, which we have addressed because we found it very relevant. It is contained in H.R. 820 as a separate provision. It is not a high-performance computing application. It is in S. 4, though, as a high-performance computing application, which is development of applications using information technologies to assist manufacturing. There is, in H.R. 820, the development of a manufacturing outreach program that will have as part of it an advanced communications network to help link small- and medium-sized manufacturers and give them assistance, because we found that even though there are manufacturing centers around the country, small business owners can't take the time to travel to them to get assistance. So we are going to try to link them up electronically, in a kind of extension network. . . .

One last comment on financing. In H.R. 820 we have a Title III, which essentially provides for the government to set up some venture capital mechanisms, funneling money into other organizations that will hopefully provide venture capital financing for the development of critical technologies. There are a lot of communication technologies on the critical technologies list.

The administration hadn't accepted that. They agreed to make it a study. But at the same time, when you look at the *National Performance Review*, Sally Katzen, when she testified before us, was talking about a program to finance the development of information technology applications for government purposes through a venture capital fund. I think it was 1 percent of the agency's operating budget.

Then you have venture capital provisions in the seed act that was passed a couple of years ago, which has the government in the position of giving government money to finance venture capital, start-up funds in Eastern Europe but not for American companies doing the same thing in the United States. What we have heard is that the Treasury Department has decided to set up an interagency group to look at the whole issue of government financing for new technologies.

BRIAN KAHIN: I think Joel Engel's example of the VCR was a bad one, because the VCR business took off on the basis of off-air taping, rather than the existence of cassettes. And I think it is a problem with the PC analogy, too.

WILLIAM GILLIS: Maybe I can help just a bit. I introduced the VHS-brand VCR in August of 1977, and in fact, it did take off immediately. There was a predecessor, the Betamax, the one-hour machine that was out in 1975. We introduced the VHS in 1977 as a time-shift mechanism and to give the ability to record off the air. From the day that we first introduced it, I went into backlog for the next year and a half because of the incredible demand.

One story that I will share with you highlights the point. We were bold enough to put my name—it was an error—in one of the ads that we ran. I got a call from a lady who said, "Are you Mr. Gillis? Are you the one that is responsible for this VHS brand VCR?" I said, "Yes." She said, "Well, I called to say thank you." For the first time in 2 years her husband had taken her out to dinner on a Monday night because he was able to record Monday night football.

Then she said, "I have a problem. Last Monday night's game went into overtime and he hasn't watched it all yet and we are out of blank tapes. Could you please find me a way to get blank tapes?" And that was only one of many such calls.

Clearly, the overwhelming evidence—and there is certainly data available—is that off-air recording was the driver; you can record when you are not at home or record when a competing program is running. That information is very definitive.

BREEDEN: What was the cost of that device when it was introduced in 1977?

GILLIS: We introduced the first one at $1,000 suggested retail. We sold it to distributors for $750. We made money the first year.

MICHAEL ROBERTS: I want to make a brief comment on the issue of winners and losers and the federal role in that, which gets everybody's blood moving. I was surprised, with all of the economists in the workshop, that nobody has mentioned Christopher Ferrell. One of his observations is that one of the great advantages of capitalism comes through innovations and asset reallocation. It harnesses a lot of creative energy and improves society.

But intrinsic to that is the notion of whether an asset reallocation is going to destroy industries. When diesel engines did away with steam engines, or Henry Ford did away with horses, thousands of jobs were destroyed, but that wasn't considered a big government policy issue. Since World War II, the government has become the majority force in technological innovations. So, the government is now linked to the destruction as well as the creation of a tremendous number of jobs, which means that there isn't any way for us to get the winners and losers out of this policy space.

I think that the challenge that we have is evolving a policy space around the NII [national information infrastructure], which we really haven't done—certainly this day and a half has shown how primitive the policy space is around the NII. We have to manage that issue, not wish hopefully that it will go away.

KAHIN: There is also the example, on the other side, of the Internet—network externalities, the government leveraging with an extremely small amount of money what has turned out to be an outstanding and increasingly commercial success.

The general information infrastructure development problem has been well phrased. There are many arguments for government investment. A couple Wally Baer mentioned that haven't been pointed out are the information cost and the transaction costs, which is why the government investment of the producer's subsidy into the backbone has made sense at the beginning.

The basic problem is the information management problem and the risk of the government interfering in private investment. If the government doesn't have the information flowing to it in a way that enables it to refrain from stepping on private investment, it is not going to work. And that is a problem in the applications area, too, that we really haven't faced yet.

I was struck by Bill Gillis's comments that he has been relying on patents in this piece of the infrastructure he is building. This is a situation where the government is directly involved in issuing patents, and it is an information management problem. The government is doing an absolutely crummy job of handling patents in the software area. It is a disaster. It is taking 32 months to process the applications, and the quality control is absolutely abysmal. At the same time, what you have witnessed in industry in general is a moving away from proprietary positions because the users won't accept it.

So, you are building a proprietary system, you are placing your bets on a proprietary infrastructure system, and you are taking a big gamble that there won't be some kind of standard that passes you by.

GILLIS: On that, I think you misunderstood. While we do hold patents, we are not necessarily basing our system on patents as its protections. And you will find that the customers, who number among the household names for the products and services, would differ. If I left the impression that that was our sole safeguard, I want to point out that what we are going to deliver is the highest-quality service that is available. It is not an issue of proprietary software or not that is really driving our market, because we are taking the information of our various customers and we are delivering it back to them in a form that they and their customers find useful.

CAROL HENDERSON: I want to thank the National Research Council for including various user communities in this workshop, such as health care and education and libraries. To misquote a former FCC chair, this particular infrastructure is not just about selling toasters. Unlike, say, the agriculture or the transportation infrastructure, it really is the infrastructure that is related to what makes us human, the ability to communicate. And it certainly has the potential to be in the future the main way that humans communicate beyond just shouting distance to each other. As such, it will have enormous and probably unpredictable effects on society at large.

In the discussions on how we go forward, what the government role is, and how we rearrange the regulatory landscape, I think that the effects on society are something that must be considered, if it is not just economic stimulus alone. Concerning some of the government roles we talked about, there seems to be some general consensus on what might be useful, such as stimulating certain kinds of applications and fostering at least start-up activities for certain communities. Those roles are very useful, and I support them.

But in a way, I take libraries, which I represent, as an example. Giving a library a little bit of money to get started in doing something new in the telecommunications or computer networking area is very useful. It requires that institution to come up with some matching money and stimulates a lot of additional activity. But it doesn't necessarily solve the systemic problem of making sure that that publicly funded institution is able to get from the infrastructure what it needs to continue its mission in the future.

And some of what, say, schools and libraries will need is similar to what other communities need in terms of ubiquity and interoperability and a fair amount of capacity and so on. But there may be other things needed that don't necessarily coincide with the mass market approach—pricing mechanisms, for instance, such as the ability to predict what that cost is going to be and to have that cost not be too usage sensitive. Otherwise use in the education, research, and library areas will be hindered, which actually is harmful to society.

These are points that we need to pay attention to in the basic regulatory structure.

BAER: I think you raise very good points, and you bring us back to the question of pricing and appropriate regulation, which is very real. I also relate to your early comment about selling toasters because, interestingly, we are dealing with different cultures in the various media, in terms of how we are generating universal access. Obviously, in the broadcast media, the way we get near-universal access to low prices is to use advertising in support of content.

It is interesting that the culture of the Internet so far has not only avoided that kind of service, but has also actively rejected it in many cases. I think it is an issue, as we move ahead and look toward larger mass use of the Internet-related networks, as to whether we are going to allow selling toasters to be a part of the system.

ALFRED AHO: I would like to make two points that I don't think have had adequate attention during this workshop. One is the importance of software in the information infrastructure. A number of industries today, say, the telecommunications industry, run on hundreds of millions of lines of code in operation. To develop 100 million lines of code could cost anywhere from $10 to $100 per line of code. We are talking about investments of tens of billions of dollars to create the software to run the infrastructure. I just want to make sure that people understand the importance of software.

Yet, there is another consequence to this, and that is the embedded base of software that is already there. And this embedded base may be more of a bottleneck than many people realize. Companies and universities just can't afford to get rid of—or the Internet, in particular can't get rid of—its embedded base of software overnight.

Bill Gillis mentioned the importance of standards and interoperability. I would like to add a third point, and that is evolvability of the infrastructure. The technology is changing very quickly, and the applications are also changing very quickly. Whatever kind of infrastructure we put in place should be an evolvable one, so that 20 years from now we won't be stuck with a narrow-gauge railway system that we can't get out of. We have seen somewhat similar effects already with the current infrastructure that is in place.

The second observation that I want to make with respect to direct investments (and in particular, whether this is something for the government) concerns investment in people, people who understand not only how to create the new infrastructure, and the evolving infrastructure, but also how to use the infrastructure.

Vice President Gore talks about reinventing government, and we have had a number of excellent presentations about how, by changing business processes or the process we use in various segments of industry and other human endeavors in health and education, we can do business and restructure the manner in which we interact with one another and conduct business in those arenas.

I think we are just now beginning to see, in a very early stage, the profound impact of this kind of business prospect—restructuring—that is going to take place in almost every area of human endeavor because of this infrastructure. What we need to do is to train people to be able to take advantage of this.

ROBERT KAHN: I couldn't agree more about the importance of software. Between the physical infrastructures and the end-user applications is the so-called middleware, which is largely services based on software. It is, in some sense, the network analog of the operating system, plus some more things, that is all software-based. And obviously, investing in people is what you ultimately end up doing with your investments. I agree with Al Aho completely.

I would like to comment about that $5 billion number that Laura threw out, just for clarification. Basically, as I understood that number, it goes for a number of things that are unique to defense. Defense has a centralized capability for communications and also a distributed part that the Army, the Navy, and the Air Force separately administer. They have their command and control messaging systems, which are part of the command structure of the military, for getting orders out and controlling our forces. They have a command and control backbone that they maintain. It is secured and protected against a variety of threats.

They worry about security technology a lot. They worry about systems that operate on the ground, in a field headquarters of the Army, and on ships, planes, and a variety of space-based systems like MilStar and the GPS [global positioning system], and they also provide support for a lot of other organizations. The R&D involved in those systems, often unique, one-of-a-kind military systems, is large. That is what the R&D budget goes for. The actual maintenance and operation of that stuff can be much larger.

DALE HATFIELD: While we are congratulating ourselves on the success of the Internet, I think we should remember something that the FCC [Federal Communications Commission] called the enhanced service exemption, and also the special access exemption from the payment of access charges, which means, essentially, that when you make a circuit-switched call, you pay a subsidy, and when you make a packet-switched call, you do not. Obviously the packet-switching community has benefited greatly by that exemption. I think we have to be very careful when we talk about success to recognize that we are not calculating all the costs. I think that should be considered.

MESSERSCHMITT: Can you elaborate on that a little bit more, Dale?

HATFIELD: Yes. Let me give you an example. My elderly parents live in Ohio and I call them on a regular long-distance network. About half the cost of the call is split between the two carriers on either end. And a fraction of that is a subsidy that goes to maintain universal service.

When I call on the packet node in Denver and I sit there for two or three hours, it is a flat-rate call and I am paying no common carrier line charge on the originating side of that or on the terminating side. So, essentially, when I make that packet-switched call, I am not paying the subsidies that I do when I make a circuit-switched call. And that is one of the reasons that a fax call is free on the Internet and yet costs X dollars on the circuit-switched network, because of the payment for access. When I call, an enhanced service exemption of the FCC says I can make a circuit-switched call to that packet network and I am not paying the access charge. It is treated as a local call, even though it is terminating in Columbus, Ohio.

JACKSON: When the Commission proposed removing the enhanced service exemption, it became quite contentious.

HATFIELD: And we were inundated with computer-generated mail that went to Congress that put political pressure on the Commission to maintain the exemption.

JACKSON: Never pick a fight with an interest group whose hobby is word processing!

HATFIELD: Let me make one other distinction that I think we are missing here. I think we all agree that most taxpayers are also telephone subscribers, that those two communities are roughly the same.

There seems to be a consensus against government investment. But then, I hear Chuck Jackson saying that if we just speeded up depreciation a little bit . . . , and I hear Joel Engel saying that if we have all these billions of dollars of free cash flow, if you just make it a little bit more attractive, we will make this investment.

But if you are making an investment as a utility and there are constitutional provisions against confiscation and so forth, if this turns out to be a bad investment—in other words, if the investment doesn't work out—then you and I as ratepayers are going to be stuck with paying for that investment through higher rates. I think some of the distinction here between private investment and public investment is a little bit wrong, because ultimately, as ratepayers, we are going to be responsible for that investment at some point anyway, if it is made under the traditional sort of public utility type of investment.

So my conclusion is, let's not kid ourselves. There is not so much difference between public investment and private investment as one might think, from the public utility standpoint.

MESSERSCHMITT: Thank you, Dale. Any comments on that?

JACKSON: Yes, I want to respond, because there are some differences. One is that, although the budget process is difficult and contentious, once you vote to make the investment, you can go ahead.

With private-sector investment and even the regulated utilities, if their management decides that they aren't going to get that money back, then they aren't going to make that investment. We haven't seen that kind of drawing up of investment in telecommunications yet, but I understand that with electric utilities now, about half to two-thirds of the generating capacity being built in this country is what are called qualified facilities rather than being built by the utilities. The combination of prudence reviews, disallowances, nuclear plants, the whole nuclear fiasco, has dried up utility investment in power-generating facilities.

If a telephone company chooses not to invest in new infrastructure, it is very hard to order it to do more. It is possible, but it is a tough fight.

HATFIELD: On the other side of the coin, if they go ahead and do it and it turns out to be a bad investment, we are not going to let telephone companies go out of business.

JACKSON: Well, we might let them go out of business at some time in the future. We will have to wait and see. We have let long-distance carriers go out of business when they imprudently invested in obsolete satellite facilities, things like that. But if it is the only phone company in town, then I think society will be very reluctant to let it go out of business, although we have had some electric utilities lose a huge amount of money in nuclear power plants. I think that nuclear power plants lost something like 10 percent of their equity in the late 1980s from failed nuclear power plants.

I think that some people have put forward a variety of schemes, like price caps, for moving away from traditional utility regulations so that you can free up this investment and not face the specter that, if the investment fails, then the residual monopoly ratepayers are going to bear all the costs.

But those are very contentious issues, and a lot of people who support reform are quite unwilling to say okay, we will let it go, we will cap the prices, and if the returns run up to 30 percent on equity because of some great new efficiency, we will let them keep it.

JOHN RICHARDSON: I believe one of the most important points made by the panel, through Laura Breeden's remarks, was the need for an estimate of the benefits of the NII. I have felt this for a long time and I am very glad to hear this come out.

I do remember one of Bob Lucky's popular columns in *IEEE Spectrum* a few months ago, perhaps a year ago, discussing the NREN. I believe his point of view in that column was build it and they will come. Perhaps much of the NII is based on this same sort of religion, but for firm policy development, I do think that we need some sort of impetus. We need to begin that task.

My question for the panel is, Do you have some ideas or some suggestions of how to approach this estimating of the future benefits?

BAER: That is obviously a hard topic to quantify at this point, because of both the long-term nature of the investment and the problem of confounding a number of government objectives—the equity objective along with any kind of efficiency ones.

One very crude measure of the effectiveness of government programs past the R&D stage is to require matching funds from private industry. How much is the private sector willing to invest to match a dollar of government money? That is not the only measure, but it seems to me it is a reasonable one to place on a whole variety of proposed government investments beyond basic R&D. And on that measure, so far, it looks like the NREN would have come out pretty well.

RICHARDSON: You are suggesting that we feel our way. And I am asking for something more than that.

BREEDEN: I am not an economist, so I am not putting my professional reputation at stake here and I can say anything I want. Having prefaced my remarks with that comment, I would say that if you are going to make an investment decision as a nation, you want to try to evaluate the long-term or medium-term benefits of that decision.

Now, if this information infrastructure succeeds, presumably there are going to be more devices sold, which means healthier industries in computing and personal communications systems and so forth. It means that more transport is going to be sold, which I think is why you see lots of phone companies doing deals with lots of entertainment and information service providers.

I think you can probably make a case—and again, I am not an economist—that a lot of spending decisions in the private sector are going to be stimulated by the development of an infrastructure that we can all use. If we look at highways and cars, and gasoline and fast food, there is maybe a similar relationship there. And I think the exercise is worth doing, because, you know, we, as a nation, have some very hard choices to make. Are we going to increase our national debt indefinitely, or are we going to do something different with our capital and our country?

KAHN: It is clearly an interesting question to pose, and it would be very nice if we could get some answers to it. You could even broaden the question to ask not only what the benefits are, but also what the return on investment is, and you could phrase that in various different terms. The thing that I would like to caution us on is not holding our breath waiting for an answer to those questions, because I am not sure they are going to be very easily forthcoming.

If you were to try to ask that same question in other areas where, I think, people have naturally just bought into the concept, you would have an equal amount of difficulty. For example, if you asked what the benefits of personal computers are in the workplace, I can make a strong argument that they are negative as well as positive, that you are spending too much money and it takes too much time and it has encumbered you. If you were to ask about the value of education, which we spend a lot of money on, you would have a tough time quantifying the benefit of any one year of the educational process, or the whole thing, or even what society would be like if you had more or less of it or a different kind. Likewise, I think you have much the same problem with health care, despite the fact that everybody knows the need for it and wants it. Estimating the benefits of it would be very hard to do.

I think you would have an even worse time trying to explain what the benefits were of the economy in general, even though I think you could make a better stab at that than perhaps all the others. So I just wouldn't hold my breath waiting for the answer, despite the fact that I wouldn't mind seeing something that tries to describe the answer.

RICHARDSON: I appreciate your point of view and I agree with it. My point of view is to encourage somebody to try to get quantitative, even though it is a feeble first step.

BRIDGER MITCHELL: I think Laura steers us in an interesting direction of trying to be specific. But trying to answer the broad question about the benefits of the NII is almost doomed to fail from excessive generality. If we can focus on specific programs, specific kinds of initiatives, there is some chance of advancement.

For example, we have a particular quantitation of the way federal action has held up gains in consumer value in the cellular area where, for nearly 10 years, it delayed the introduction of a technology that was clearly there. We have market evaluation processes for measuring that sort of thing. But the market may not send adequate signals to evaluate the "public good" aspects of education and consumption of raw information.

CLIFFORD LYNCH: In listening to this, I have felt that an aspect is missing in some of the discussions. I want to try to quickly outline it and then solicit some reaction.

I think that when we have been talking about this information infrastructure and looking at some of the economic models that might provide insight, we tend to be thinking about a communications network. And we think about universal service objectives in terms of getting everybody connected. I have some reservations about whether the primary use of this new network world is going to be person-to-person communication for the vast majority of users. Certainly, things like the telephone work really effectively for a lot of people right now. And things like the picture phone haven't been roaring successes.

I suspect that a lot of the use of this system is going to be access to various kinds of information services. If you think of that as accounting for a lot of the use of an information infrastructure, a bunch of questions come up. What does universal service mean in that sense? You have connected people to something, but they can't afford to use most of the services on it, because they cost many many times what the connection costs. As with the phone, just because we give them universal service doesn't mean that we give everybody a blank check for 900 numbers.

Another point. It seems to me that another area of potential government investment choices to be considered is placing information services on your network, including, as you mentioned, some of the enormous stores of government information that could be organized and made accessible, either at very low cost or perhaps, in some cases, for free if there are policy or legislative mandates to do so, for access by the citizenry. It seems to me that that is a whole other dimension of potential government investment that could move along the creation of the infrastructure by giving people reasons to want to make use of it.

BREEDEN: I think Cliff is right. It is incorrect to focus only on the communications aspect of this and to assume that that means people communicating directly with other individuals. One of the other panelists made a point about the need to standardize presentation of data. Cliff has done a lot of work in something called Z39.50, which is an information standard. I think those areas are going to be tremendously important.

I tend to think that the equity issues, if we are going to avoid having a society of information "haves" and "have nots," are going to be solved in the public libraries.

MESSERSCHMITT: Laura made a very important point from my perspective—that is, increasingly, the cost to users in terms of equipment and services is often for equipment that they buy themselves, rather than for network transport services, which are rapidly being driven toward zero. Bob Lucky made this point in his talk, that Internet was so cheap in comparison to the telephone. So I think that the issues with respect to customer-owned equipment that is necessary are quite different, probably, from the issues having to do with the communication infrastructure and yet may ultimately be more important in terms of issues like universal service.

Appendix

Contributors and Participants

Alden F. Abbott
National Telecommunications and
  Information Administration

Duane A. Adams
Advanced Research Projects Agency

Alfred V. Aho
Bell Communications Research
[*currently Columbia University*]

Robert J. Aiken
U.S. Department of Energy

Walter S. Baer
RAND Corporation

Jonathan Baker
Council of Economic Advisors
Executive Office of the President

Audrey Bashkin
House Government Operations Committee
U.S. Congress

Jane Bortnick-Griffiths
Library of Congress

William Braun
Motorola Inc.

Laura L. Breeden
FARNET
[*currently National Telecommunications
  and Information Administration*]

Tim Brennan
University of Maryland

Raul G. Catangui
Corning Incorporated

Vinton G. Cerf
Corporation for National Research
  Initiatives

Nancy M. Cline
Pennsylvania State University

Mark Coblitz
Comcast Corporation

Nina W. Cornell
Private consultant

Robert W. Crandall
Brookings Institution

Colin Crook
Citicorp, N.A.

Mary Jo Deering
U.S. Department of Health and
  Human Services/Public Health Service

Stephen J. Downs
U.S. Department of Health and
  Human Services/Public Health Service

Michael Einhorn
U.S. Department of Justice

Joel Engel
Ameritech

Joseph Farrell
University of California at Berkeley

Kevin Finneran
Issues in Science and Technology

Tim Finton
Honeywell Inc.

Charles M. Firestone
Aspen Institute

Kenneth Flamm
U.S. Department of Defense

Henry Geller
The Markle Foundation

Robert Gellman
House Government Operations Committee
U.S. Congress

George Gilder
Private Consultant

William F. Gillis
Motorola Inc.

Robert G. Harris
University of California at Berkeley

Dale N. Hatfield
Hatfield Associates Inc.

Carol Henderson
American Library Association

Robert Y. Huang
TRW Inc. (retired)

Charles L. Jackson
Strategic Policy Research Inc.

Jeffrey M. Jaffe
IBM T.J. Watson Research Center

Brian Kahin
Harvard University

Robert E. Kahn
Corporation for National Research Initiatives

Thomas Kalil
National Economic Council
Executive Office of the President

George Kohl
Communications Workers of America

Alfred M. Lee
National Telecommunications and
  Information Administration

Ted Leventhal
Telecom Data Report

Donald A. Lindberg
The National Library of Medicine

Joan Lippincott
Coalition for Networked Information

Thomas J. Long
Toward Utility Rate Normalization (TURN)

Edward D. Lowry
Bell Atlantic Corporation

Robert W. Lucky
Bell Communications Research

Clifford A. Lynch
University of California,
  Office of the President

Bruce W. McConnell
Office of Management and Budget
Executive Office of the President

Lee McKnight
Massachusetts Institute of Technology

James Mecklenburger
The Mecklenburger Group

David G. Messerschmitt
University of California at Berkeley

Brady Metheny
Washington FAX

Bridger M. Mitchell
RAND Corporation
[currently Charles River Associates]

Rex Mitchell
Pacific Bell

Paul Mockapetris
Advanced Research Projects Agency
*[currently University of Southern California,
Information Sciences Institute]*

Michael R. Nelson
Office of Science and Technology Policy
Executive Office of the President

David Nicoll
National Cable Television Association

Eli M. Noam
Columbia University

Roger G. Noll
Stanford University

Charles Oliver
National Telecommunications and
Information Administration

Kevin Patrick
U.S. Department of Health and
Human Services/Public Health Service

Robert Pearlman
Private Consultant

Robert Pepper
Federal Communications Commission

Lucy Richards
House Science, Space, and
Technology Committee
U.S. Congress
[currently U.S. Department of Commerce]

John M. Richardson
University of Maryland

Linda Roberts
U.S. Department of Education

Michael M. Roberts
EDUCOM

Glen Robinson
University of Virginia Law School

Jim Schubener
New Technology Week

Gail Garfield Schwartz
Teleport Communications Group

Ali Shadman
Ameritech

Scott Shenker
Xerox PARC

Edward H. Shortliffe
Stanford University School of Medicine

Terri A. Southwick
National Telecommunications and
Information Administration
[currently Patent and Trademark Office]

Raymond L. Strassburger
Northern Telecom Inc.

Myron Struck
Communications Daily

Michael Telson
House Budget Committee
U.S. Congress

Suzanne P. Tichenor
Council on Competitiveness

Keith Townsend
Sprint Inc.

George L. Turin
Teknekron Corporation

Elana Varon
Federal Computer Week

Philip L. Verveer
Willkie, Farr, and Gallagher

Philip Webre
Congressional Budget Office
U.S. Congress

Charles Wessner
Private Consultant